Nanomaterials - Toxicity, Human Health and Environment

Edited by Simona Clichici, Adriana Filip and Gustavo M. do Nascimento

Published in London, United Kingdom

IntechOpen

Supporting open minds since 2005

Nanomaterials - Toxicity, Human Health and Environment
http://dx.doi.org/10.5772/intechopen.78085
Edited by Simona Clichici, Adriana Filip and Gustavo M. do Nascimento

Contributors
Paulo Dias, Daniela Sousa Coelho, S Kalaiselvan, Manivannan J, Josef Mašek, Eliška Mašková, Daniela Lubasová, Jaroslav Turanek, Roman Špánek, Milan Raška, Akeel M. Kadim, Xiaoniu Yu, Qiwei Zhan, Gustavo Morari Do Nascimento, Guilherme Fredeico Bernardo Lenz E Silva, Camila Viana, Fernanda Vieira, Danieli Silva Domingues, Daniela-Rodica Mitrea, Alina-Mihaela Toader, Oana-Alina Hoteiuc, Matheus Mantuanelli Roberto, Cintya Christofoletti, Lavinia Gabriela Carpen, Tomy Acsente, Maria Adriana Acasandrei, Elena Matei, Claudia Gabriela Chilom, Gheorghe Dinescu, Maria Zaharescu, Jorge Alberto Sanchez-Burgos, Fabián Razura-Carmona, Alejandro Pérez-Larios, Marco Vinicio Ramírez-Mares, Mayra Herrera-Martínez, Gladys A. Prado Guzmán

First published in London, United Kingdom, 2020 by IntechOpen
IntechOpen is the global imprint of INTECHOPEN LIMITED, registered in England and Wales, registration number: 11086078, 7th floor, 10 Lower Thames Street, London, EC3R 6AF, United Kingdom
Printed in Croatia

British Library Cataloguing-in-Publication Data
A catalogue record for this book is available from the British Library

Additional hard and PDF copies can be obtained from orders@intechopen.com

Nanomaterials - Toxicity, Human Health and Environment
Edited by Simona Clichici, Adriana Filip and Gustavo M. do Nascimento
p. cm.
Print ISBN 978-1-78984-616-4
Online ISBN 978-1-78984-617-1
eBook (PDF) ISBN 978-1-78984-329-3

For EU product safety concerns:
IN TECH d.o.o., Prolaz Marije Krucifikse Kozulić 3, 51000 Rijeka, Croatia, info@intechopen.com or visit our website at intechopen.com.

Meet the editors

Simona Clichici has her PhD in Medicine and is a Professor of Physiology at "Iuliu Hatieganu" University of Medicine and Pharmacy Cluj-Napoca, Romania. In 2007, she became the coordinator of the in vivo Research Laboratory at the Physiology Department. She has practical experience and many published papers concerning in vivo experimental design of different pathological conditions (e.g. liver fibrosis, diabetes mellitus), oxidative stress and cytotoxicity assessment, study of the nanomaterials' effects on living organisms (cytotoxicity, oxidative stress, inflammation, apoptosis), protective effects of different natural compounds, photodynamic therapy, and carcinogenesis. She was the coordinator of two national grants, one of them regarding the toxicity of carbon nanotubes and the author/co-author of more than 100 papers, all of them concerning experimental studies on laboratory animals.

Adriana Filip has been a Professor at the Physiology Department at "Iuliu Hatieganu" University of Medicine and Pharmacy Cluj-Napoca, Romania.since 2017 and has her PhD in medicine since 2004. She is a coordinator of the *in vitro* Lab since 2009 and a founding member and treasurer of ASOM – Association of Oxidative Stress in Medicine and BIOCOM - Biocompatibility Study Center for Natural and Synthetic Products. She is the director of two national projects and responsible partner of one project; author/co-author of 71 ISI articles (43 as the main author and 28 as the co-author), IF cumulative as main author = 115.53; Hirsch index = 14; http://www.researcherid.com/rid/C-2906-2011. Her areas of scientific interest include: the photochemopreventive effect of natural compounds in skin tumors, the anti-inflammatory and anti-proliferative effect of gold and silver nanoparticles and signaling pathways involved, cytotoxicity of synthetic compounds and natural extracts on fibroblasts and stem cells from the dental follicle, and photodynamic therapy in experimental tumors.

Dr. Gustavo Morari do Nascimento is a professor at the Federal University of ABC. He has experience in many fields related to the characterization of nanomaterials by using spectroscopic techniques associated with microscopic resolution. He obtained a doctoral degree from the University of São Paulo in 2004 with a thesis on spectroscopic characterization of nanocomposites formed by conducting polymers and clays. Subsequently, in 2007/2008, he gained a post-doctoral degree at Massachusetts Institute of Technology on the resonance Raman study of double-walled carbon nanotubes doped with halogens under the guidance of the legendary Mildred S. Dresselhaus. Back in Brazil, he spent three years (2009–2011) at the Federal University of Minas Gerais working with the synthesis of nanostructured carbon modified with molecular magnets. Nowadays, his research focus is on molecular characterization of modified carbon nanostructured materials and polymer nanocomposites using different spectroscopic techniques. Resonance Raman and surface-enhanced Raman spectroscopy coupled to microscopy techniques added to X-ray absorption techniques at the National Synchrotron Light Laboratory were the main techniques employed in the investigation.

Contents

Preface

In the last decade, nanomaterials have become a double-edged sword. On the one hand, there has been an increase in the production of nanomaterials, since they have proven their limitless potential not only for technological applications, but also for medical ones. On the other hand, the increasing use of these nanomaterials has raised concerns regarding their safety for environmental and human health, due to their potential toxicity. Their faith in the human body, along with their interactions with different tissues, became of vital importance. The toxic effects of nanomaterials depend on their type, synthesis, surface geometry, diameter, length, and functionalization.

This book discusses the main and new aspects related to nanomaterials' toxicity divided into four main areas: Assessment of nanomaterials' toxicity; Environmental and health impact of nanomaterials; Modulation of nanomaterials' toxicity; Characterization and applications of nanomaterials.

The overall idea of the book is to provide the reader with an evidence–based, comprehensive, and up-to-date overview of the current state of the art of nanomaterials' toxicity, including their synthesis and characterization, environmental impact, tests to assess their toxicity *in vitro* and *in vivo*, ways to modulate their impact on living organisms, and their beneficial use in biomedical applications.

At the beginning of the book, Professor Gustavo Nascimento presents an overview of a challenging up-to-date subject, nanofibers. In the Prologue, the main existing results regarding the synthesis, characterization through Raman spectroscopy, and applications of polyaniline nanofibers are reviewed.

In Chapter 2 the authors discuss the most recommended and frequently used methods, including both *in vivo* and *in vitro* tests, which are the most suitable ones for the assessment of nanomaterials' toxicity. Also, toxicity of the environment is taken into account, along with toxicity related to human health. Finally, the need for the standardization of these methods is discussed.

Chapter 3 focuses on carbon nanomaterials and presents new tools and methodologies to assess the exposure and risk evaluation of hazards used in health, safety, and environmental management of these kinds of structures. Possible relations between safety aspects and biokinetics interaction of living organisms according to the exposure route, along with the major protocols, standards, and guidelines on the safe handling of nanomaterials, are also presented.

The environmental impact of urban atmospheric nanoparticles is presented in Chapter 4, with a focus on the oxidative stress produced by this type of nanomaterial. The authors discuss different mechanisms involved in the interaction between these nanomaterials and living organisms, including cellular internalization, activation of signaling pathways, decrease of cellular antioxidants, activation of the proinflammatory cascade, lipid peroxidation, and activation of the cellular signaling pathway leading to apoptosis, with impacts on human health.

The toxicity of tungsten nanoparticles with respect to normal human skin fibroblast cells is evaluated in Chapter 5. The authors have analyzed the cytotoxic effects of tokamak dust produced in laboratory cell lines on human fibroblasts and have demonstrated that the effects are dose dependent. At low concentrations (<100 µg/mL), tungsten nanoparticles proved to have no toxic effects, while at concentrations up to 2 mg/mL they can exert toxic effects on the cells.

Chapter 6 focuses on the methods used for the synthesis of nanomaterials, methods by which shapes and sizes could be controlled, and also looks at the methods used to characterize the biomaterials. The manipulation of nanomaterials' shape and size could lead to less toxic nanomaterials.

A new method for the treatment of heavy metal ions in soil or water is presented in Chapter 7. Heavy metal ions can be mineralized by phosphate-mineralization microbes and stable phosphate nanomaterials are formed. This is a mineralization method that can remove heavy metal pollutants from soil or water. In addition, heavy metal pollution can degrade the environment and be a threat to human health.

Chapter 8 focuses on the use of SiO_2-based nanomaterials as support for compounds with biological activity such as antibodies and enzymes. The authors emphasize the synthesis of SiO_2 nanomaterials with different morphologies, their physicochemical characteristics, the biocatalytic activity of immobilized enzymes on simple SiO_2, and their behavior dependent on the morphology of SiO_2 inorganic carriers obtained by the sol-gel method.

Chapter 9 is a revision of the density function theory calculation methods of the vibrational zero point for organic molecules containing silicon atoms. This class of molecules are the building blocks of nanomaterials. In addition, the authors provide an extensive amount of calculated data compared to other literature. The chapter gives abundant material for the spectroscopic characterization of advanced materials.

Chapter 10 gives a detailed calculation of the vibrational modes, proton and carbon nuclear magnetic resonance spectra, and other different properties for 5-nitro-1,3-benzodioxole and derivatives. The non-linear optical behavior of the examined molecule is investigated by the determination of hyperpolarizability. This result indicates that 5-nitro-1,3-benzodioxole is a good candidate for non-linear optical study.

A broad review of the synthesis and characterization of carbon nanotubes is described in Chapter 11. Since their discovery, carbon nanotubes have offered tremendous opportunities for the development of new materials and composites. This chapter provides a report on recent advances in the science of carbon nanotubes and their potential applications.

Chapter 12 focuses on a number of applications of cadmium telluride quantum dots (CdTe QDs) in nanotechnology. A hybrid device fabricated from three layers—an organic polymer, an electron injector from organic molecules, and a semiconductor material (CdTe QDs)—proved to be effective in white light generation. The synthesis, characterization, and evaluation of optical, electrical, morphological, and electroluminescent properties of CdTe QDs and of the obtained device are presented and discussed by the author.

Even though the knowledge regarding the real impact of nanomaterials on the environment and human health is still limited, we hope that this book will provide a useful tool for the reader to understand the complex field of nanotechnology and nanomedicine.

Dr. Simona Clichici and Dr. Adriana Filip
Professor,
Department of Physiology,
University of Medicine and Pharmacy,
Cluj-Napoca, Romania

Dr. Gustavo M. do Nascimento
Professor,
Centre for Natural Sciences and Humanities,
Federal University of ABC,
Santo André, Brazil

Section 1

Prologue

Chapter 1

Prologue: Nanofibers

Gustavo M. Do Nascimento

1. Introduction

The preparation of polymers with morphology well determined in the nanometric range is one of the great challenges in the polymer science and technology. The possibility to prepare nanofibers (or nanofibers) brings the opportunity to produce polymers with new or reinforced properties. Many ways have been developed to synthesize polymeric nanofibers, for instance, the polymerization into media having large organic acids. The interfacial polymerization can also form nanofibers at an aqueous-organic interface. Hence, a great variety of "bottom-up" approaches, such as electrospinning, interfacial, seeding, and micellar, can be employed to obtain pure polymeric nanofibers. The preparation of nanostructured polymers by self-assembly with reduced post-synthesis processing warrants further applications, especially in the field of biotechnology and removable resources. The notable applications include tissue engineering, biosensors, filtration, wound dressings, drug delivery, and enzyme immobilization. In this chapter, the state-of-the-art results of synthesis, spectroscopic characterization, and applications of polyaniline nanofibers will be reviewed. The main goal of this work is to contribute to the rationalization of some important results obtained in this wonder area of polymeric nanofibers.

2. Nanofibers

Despite that nanofibers are produced for a long time, only in recent years, the scientific interest in this field has rapidly increased. The reason for that is, probably, owing to the improvement of the synthetic pathways in the production of better nanofibers. In addition, the combination of spectroscopic and microscopic techniques leads to a better corrletion between structure and properties of nanofibers. **Figure 1** shows that in 2018, more than 6000 papers having "nanofiber" or "nanofibre" as keyword were published. In addition, **Figure 2** shows that at least 20 different research fields have more than 1000 papers published related to "nanofiber" or "nanofibre." These two graphs clearly show that nanofibers are one of the focuses in the science of advanced materials.

Our group has dedicated to the preparation and characterization of polyaniline nanofibers [1–10]. Among the different techniques used for structural investigation, resonance Raman spectroscopy is the most important technique for these systems. Thus, in this chapter, mainly the Raman results obtained for polyaniline (PANI) will be discussed.

3. Nanofibers of conductive polymers

Nowadays, the preparation of conductive polymers with organized morphology and structure is a desired deal. Since the discovery of poly(acetylene) doping process

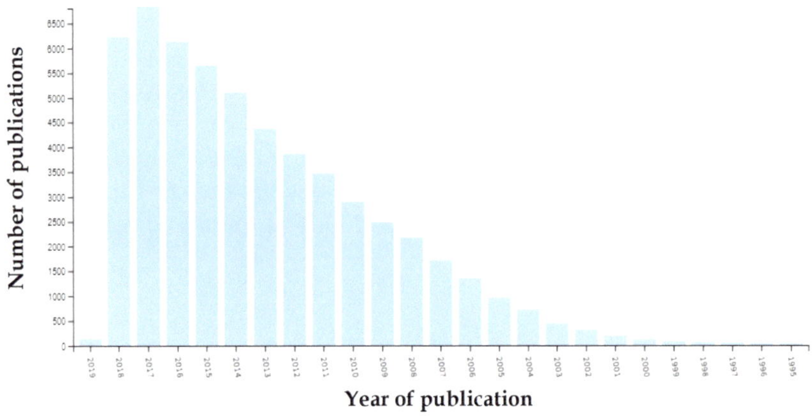

Figure 1.
Number of publications by year having the keyword "nanofiber" or "nanofibre" in the text. The research was done in November 25, 2018, using Web of Science database. The total score found are 54,611 papers.

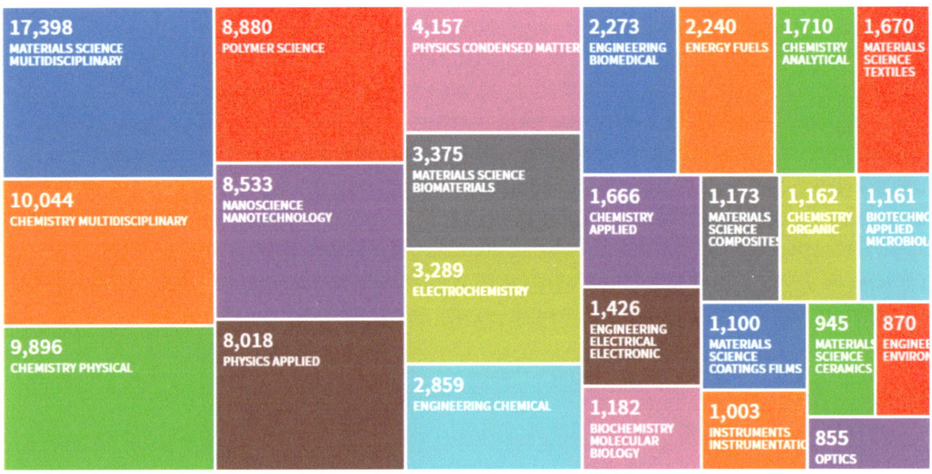

Figure 2.
Number of publications by year having the keyword "nanofiber" or "nanofibre" in the text divided by the main research areas or categories. The research was done in November 25, 2018, by using Web of Science database.

in the early 1970s [11–16] and posterior investigation of its properties mainly done by Hideki Shirakawa, Alan J. Heeger, and Alan G. MacDiarmid (see **Figure 3**), the field of conductive polymers brings many contributions to different applications: from batteries to organic light-emitting diode (OLED) displays. The preparation of nano-structured conductive polymers can turn the polymer more efficiently to applications. The doping process [17–25] in conjugated polymers is characterized by the passage from an insulating or semiconducting state with low conductivity, typically ranging from 10^{-10} to 10^{-5} Scm^{-1}, to a "metallic" regime (ca. 1–10^{4} Scm^{-1}; see **Figure 3**).

Reversibility is one main characteristic of chemical doping; in fact, the polymer can return to its original state without major changes in its structure. Counterions stabilize the doped state in the polymeric chain. The conductivity can be modulated only by adjusting the doping level, varying from non-doped insulating state to highly doped or metallic. All conductive polymers (and their derivatives), for example, among others, may be doped by p (oxidation) or n (reduction) through chemical and/or electrochemical process [16–18]. The doping process can also be characterized by no loss or gain of electrons from external

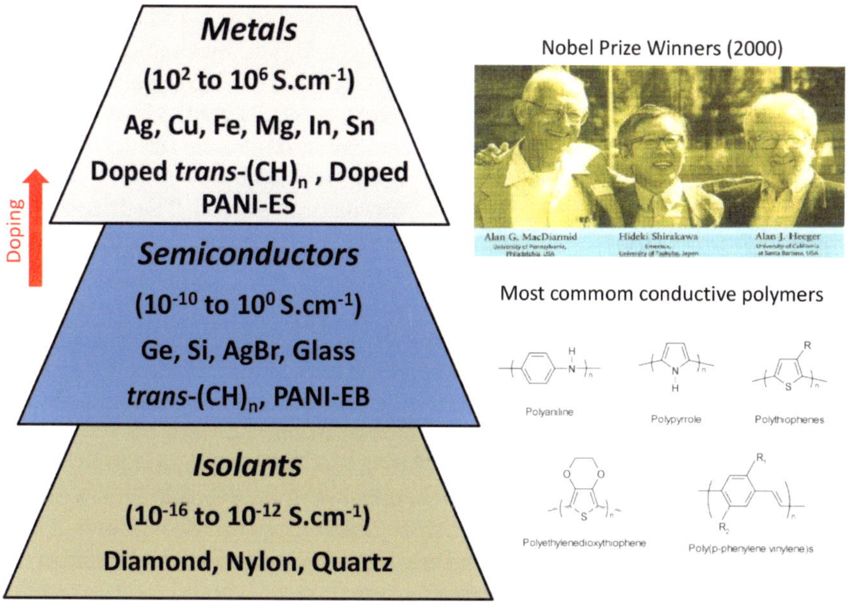

Figure 3.
The Nobel winners (Hideki Shirakawa, Alan J. Heeger, and Alan G. MacDiarmid) and the chemical structures of the most common conductive polymers. The conductivity values for different materials are displayed in comparison with conducting polymers before and after the doping process. The doping causes (addition of nonstoichiometric chemical species in quantities commonly low ≤10%) dramatic changes in the electronic, electrical, magnetic, optical, and structural properties of the polymer.

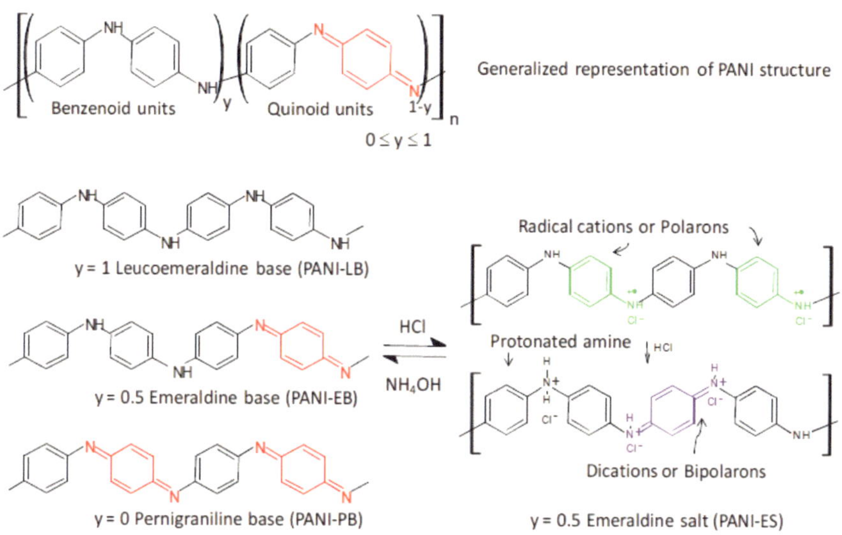

Figure 4.
Generalized representation of chemical structure of PANI and its most common forms.

agents. This is the point for polyanilines (see **Figure 4**), and this process is named internal redox process.

PANI-ES is formed after protonation with the appearance of the free radical tail of band in the NIR spectral region (starting from ca. 1.6 eV or 780 nm), which is attributed to a charge transfer from the highest occupied energy level of the benzene ring (HOMO) to the lowest unoccupied energy level of a semiquinone (polarons) ring (LUMO) [25].

PANI nanofibers can be prepared by using different routes, and the resulting polymer shows improvement in its electrical, thermal, and mechanical stabilities. The conventional synthesis of polyaniline, based on the oxidative polymerization of aniline in the presence of a strong acid dopant, typically results in an irregular granular morphology with a very small percentage of nanoscale fibers. Highly uniform PANI nanofibers with diameter ranging from 30 to 120 nm, depending on the dopant, are prepared by interfacial polymerization [26, 27]. The diffusion of the formed product from the interfacial solvent-solvent region to the bulk of the solvent can suppress uncontrolled polymer growth by isolating the fibers from the excess of reagents. In fact, the addition of certain surfactants to such an interfacial system grants further control over the diameter of the nanofibers. Isolation of the nanostructured PANI from the solution can be achieved by filtration in a nanoporous filters or dialyzed, and then the cleaned solution containing the nanofibers is centrifuged in order to separate the nanofibers from the solution.

Another approach is the synthesis of PANI nanofibers or nanotubes by making use of large organic acids. These acids form micelles upon which aniline is polymerized and doped. Fiber with diameters from 30 to 60 nm can be modulated by reagent ratios [28–31]. PANI nanofibers can also be obtained in ionic liquids (ILs) as synthetic media [2, 6]. There is a large variety of ionic liquids, and the most used ones are derived from imidazolium ring, pyridinium ring, quaternary ammonium, and tertiary phosphonium cations. The most unusual characteristic of these systems is that, although they are liquids, they present structural organization and can act as a template-like system, and PANI nanofibers are obtained when the aniline is polymerized in these media.

4. Raman spectroscopy of polyaniline nanofibers

Raman spectroscopy is a technique par excellence for probing the vibrational frequencies by inelastic scattering the incident light (see **Figure 5**) [32–35]. In the conventional Raman spectroscopy, the intensities of the Raman bands are linearly proportional to the intensity of the incident light and proportional to the square of the polarizability tensor. However, when the laser line falls within the region of a permitted electronic transition, the Raman bands that are tightly coupled or associated with the excited electronic state have a tremendous increase of about 10^{5-6} times; this is what characterizes the resonance Raman effect. In the case of multi-chromophoric system, like polyaniline, just by tuning an appropriate laser radiation on an electronic transition of the polymer, the spectrum changes dramatically (see **Figure 6**).

PANI shows a characteristic Raman bands for each oxidized or protonated form [36–40]. The presence of a free carrier tail absorption in the UV–VIS–NIR spectra for both PANI nanofibers/nanotubes prepared with NSA (β-naphthalenesulfonic acid) or with DBSA (dodecybenzenesulfonic acid) confirmed that polymeric chains have an extended conformation. In addition, the band at 609 cm^{-1} is sensible to conformation changes of the PANI chains [1, 3]. The studies of doping and heating behavior of PANI-NSA nanofibers show the loss of the fibrous morphology of PANI after treatment with HCl solution [4]. However, the PANI nanofibers are more susceptible to cross-linking (bands at 578 and 1340 cm^{-1}; see **Figure 6**) than conventional PANI, and after heating at 200°C, it is possible to dope the polymer with HCl and maintain the nanostructured morphology.

PANI nanofibers prepared from interfacial polymerization were also characterized by Raman spectroscopy. Bands at 200 and 296 cm^{-1} related to C_{ring}-N-C_{ring} deformation and lattice modes of polaron segments of PANI practically disappear in the Raman spectra of PANI nanofibers. The changes indicate the increase of the

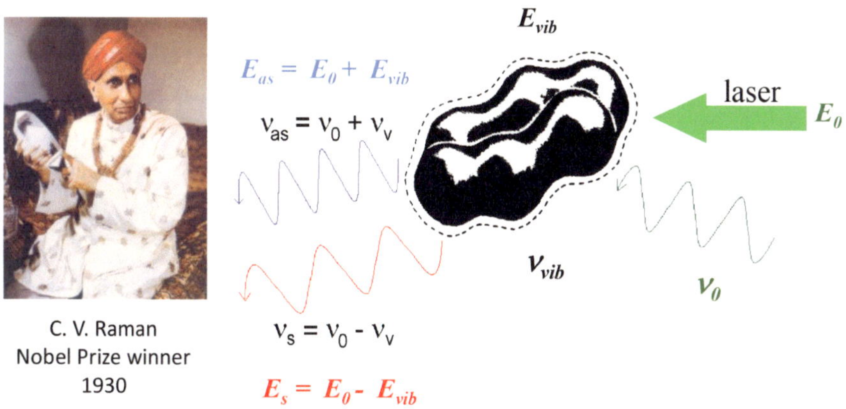

Figure 5.
Schematic representation of Raman effect. The Raman scattering was discovered by C. V. Raman and is characterized by inelastic scattering of the incident radiation (ν_o) with laser energy (E_O). The scattered light has two components: Stokes radiation (ν_s) with lower energy than E_o ($E_s < E_o$) and the anti-stoke radiation (ν_{as}) with higher energy than E_o ($E_{as} > E_o$).

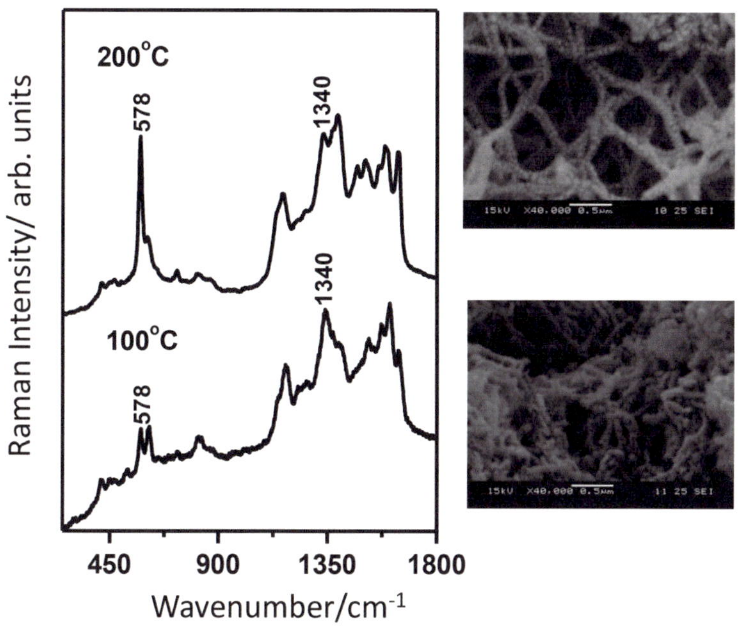

Figure 6.
Resonance Raman spectra of PANI-NSA after heating at indicated temperatures and doping with HCl. For comparison the SEM images are also given.

torsion angles of the C_{ring}-N-C_{ring} segments. In addition, the FTIR spectra for PANI nanofibers display higher changes in the region from 2000 to 4000 cm^{-1}. Both data are associated to the formation of bipolarons (protonated, spinless units) in the PANI nanofiber backbone higher than the conventional PANI. The PANI nanofiber morphology permits major diffusion of the ions inside the polymeric matrix leading to a more effective protonation of the polymeric chain [5]. In addition, only for PANI nanofibers with a diameter of 30.0 nm, low dispersion of the νC=N band is seen (see **Figure 7**). The Raman dispersion is associated to the electron–phonon coupling into a conjugated structure. In other words, very low D values indicated more electronic homogeneity into the PANI nanofibers, due to the stacking of quinoid-quinoid rings, leading to high torsion C_{ring}-N-C_{ring} angles.

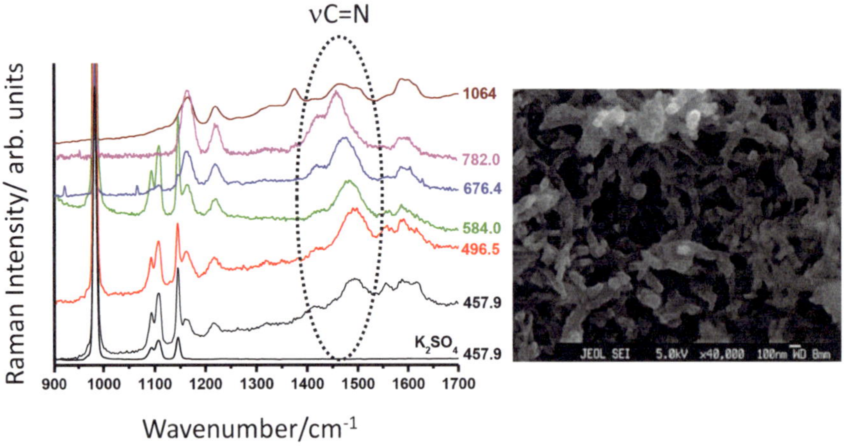

Figure 7.
Raman dispersion of PANI nanofibers.

5. Conclusion

The structural studies of the polyaniline nanofibers by using resonance Raman spectroscopy, as the main technique, have been decisive to elucidate intra- and interchain interactions and chemical and thermal stabilities of PANI nanofibers. The presence of phenoxazine rings is observed in PANI nanofibers formed in micellar media. The presence of these rings is crucial for stacking and stabilization of the fibers. In addition, the changes in bands at low energies are associated with an increase in the torsion angles of C_{ring}-N-C_{ring} segments due to the formation of bipolarons (protonated, spinless units) in the PANI nanofibers. The major diffusion of the ions inside the nanofiber gives a more effective protonation. However, only with the previous thermal treatment, it is possible to retain the nanofiber morphology.

Hence, the π-stacking between quinoid rings and the presence of π-π stacking formed by phenoxazine rings can be the driving forces for the formation of the fiber morphology of PANI. The quality of the PANI nanofibers can be monitored by the influence over the Raman dispersion curves. Finally, the example of characterization of PANI nanofibers by using Raman spectroscopy can be applied to other nanofiber materials with the improvement of future nanofiber structural studies.

Author details

Gustavo M. Do Nascimento
Federal University of ABC-CCNH, Brazil

*Address all correspondence to: gustavo.morari@ufabc.edu.br

References

[1] Do Nascimento GM, Silva CHB, Temperini MLA. Electronic structure and doping behavior of PANI-NSA nanofibers investigated by resonance raman spectroscopy. Macromolecular Rapid Communications. 2006;**27**:255

[2] Rodrigues F, Do Nascimento GM, Santos PS. Dissolution and doping of polyaniline emeraldine base in imidazolium ionic liquids investigated by spectroscopic techniques. Macromolecular Rapid Communications. 2007;**28**:666

[3] Do Nascimento GM, Silva CHB, Izumi CMS, Temperini MLA. The role of cross-linking structures to the formation of one-dimensional nano-organized polyaniline and their Raman fingerprint. Spectrochimica Acta Part A. 2008;**71**:869

[4] Do Nascimento GM, Silva CHB, Temperini MLA. Spectroscopic characterization of the structural changes of polyaniline nanofibers after heating. Polymer Degradation and Stability. 2008;**93**:291

[5] Do Nascimento GM, Kobata PYG, Temperini MLA. Structural and vibrational characterization of polyaniline nanofibers prepared from interfacial polymerization. The Journal of Physical Chemistry. B. 2008;**112**:11551

[6] Rodrigues F, Do Nascimento GM, Santos PS. Studies of ionic liquid solutions by soft X-ray absorption spectroscopy. Journal of Electron Spectroscopy and Related Phenomena. 2007;**155**:148

[7] Do Nascimento GM. In: Kumar A, editor. (Org.)Nanofibers. 1st ed. Austria/Croacia: InTech; 2010

[8] Do Nascimento GM. X-ray absorption spectroscopy of nanostructured polyanilines. Chemical Papers. 2013;**67**:933

[9] Do Nascimento GM. In: Michaelson L, editor. (Org.)Advances in Conducting Polymers Research. 1st ed. New York: Nova Publishers; 2014

[10] Do Nascimento GM. Raman dispersion in polyaniline nanofibers. Vibrational Spectroscopy. 2017;**90**:89

[11] Shirakawa H, Ikeda S. Infrared spectra of poly(acetylene). Polymer Journal. 1971;**2**:231

[12] Shirakawa H, Ikeda S. Cyclotrimerization of acetylene by tris(acetylacetonato)titanium(III)-diethylaluminum chloride system. Journal of Polymer Science. 1974;**12**:929

[13] Chiang CK, Druy MA, Gau SC, Heeger AJ, Louis EJ, MacDiarmid AG, et al. Synthesis of highly conducting films of derivatives of polyacetylene (CH)x. Journal of the American Chemical Society. 1978;**100**:1013

[14] Chiang CK, Fincher CR Jr, Park YW, Heeger AJ, Shirakawa H, Louis EJ, et al. Electrical-conductivity in doped polyacetylene. Physical Review Letters. 1977;**39**:1098

[15] Shirakawa H. The discovery of polyacetylene film: The dawning of an era of conducting polymers (Nobel Lecture). Angewandte Chemie, International Edition. 2001;**40**:2575

[16] Shirakawa H, Louis EJ, MacDiarmid AG, Chiang CK, Heeger AJ. Synthesis of electrically conducting organic polymers-halogen derivatives of polyacetylene, (CH)X. Journal of the Chemical Society, Chemical Communications. 1977;**16**:578

[17] MacDiarmid AG. "Synthetic Metals": A novel role for organic

polymers (Nobel Lecture). Angewandte Chemie, International Edition. 2001;**40**:2581

[18] Nigrey PJ, MacDiarmid AG, Heeger AJ. Electrochemistry of polyacetylene, (CH)X- Electrochemical doping of (CH)X films to the metallic state. Journal of the Chemical Society, Chemical Communications. 1979;**14**:594

[19] Han CC, Elsenbaumer RL. Protonic acids- generally applicable dopants for conducting polymers. Synthetic Metals. 1989;**30**(1):123

[20] Heeger AJ. Semiconducting and metallic polymers: The fourth generation of polymeric materials (Nobel Lecture). Angewandte Chemie, International Edition. 2001;**40**:2591

[21] MacDiarmid AG, Epstein AJ. Polyanilines- A novel class of conducting polymers. Faraday Discussions of the Chemical Society. 1989;**88**:317

[22] MacDiarmid AG, Epstein AJ. Conducting Polymers, Emerging Technologies. New Jersey: Technical Insights; 1989. p. 27

[23] MacDiarmid AG, Chiang JC, Richter AF, Sonosiri NLD. In: Alcácer L, editor. Conducting Polymers. Dordrecht: Reidel Publications; 1989

[24] MacDiarmid AG, Epstein AJ. In: Prasad PN, editor. Frontiers of Polymers and Advanced Materials. Vol. 251. New York: Plenum Press; 1994

[25] Huang WS, MacDiarmid AG. Optical properties of polyaniline. Polymer. 1993;**34**:1833

[26] Huang J, Kaner RB. Nanofiber Formation in the chemical polymerization of aniline: A mechanistic study. Angewandte Chemie, International Edition. 2004;**43**:5817

[27] Huang J, Kaner RB. A general chemical route to polyaniline nanofibers. Journal of the American Chemical Society. 2004;**126**:851

[28] Zhang ZM, Wei ZX, Wan MX. Nanostructures of polyaniline doped with inorganic acids. Macromolecules. 2002;**35**:5937

[29] Qiu HJ, Wan MX, Matthews B, Dai LM. Conducting polyaniline nanotubes by template-free polymerization. Macromolecules. 2001;**34**:675

[30] Wei ZX, Wan MX. Hollow microspheres of polyaniline synthesized with an aniline emulsion template. Advanced Materials. 2002;**14**:1314

[31] Gao H, Jiang T, Han B, Wang Y, Du J, Liu Z, et al. Aqueous/ionic liquid interfacial polymerization for preparing polyaniline nanoparticles. Polymer. 2004;**45**:3017

[32] Batchelder DN. In: Brässler H, editor. Optical Techniques to Characterize Polymer Systems. Amsterdam: Elsevier; 1987

[33] Batchelder DN, Bloor D. Advances in Infrared and Raman Spectroscopy. London: Wiley-Heyden; 1984

[34] Clark JH, Dines TJ. Resonance raman spectroscopy, and its application to inorganic chemistry. New analytical methods (27). Angewandte Chemie (International Ed. in English). 1986;**25**:131

[35] McHale JL. Molecular Spectroscopy. US: Prentice-Hall; 1999

[36] Sariciftci NS, Bartonek M, Kuzmany H, Neugebauer H, Neckel A. Analysis of various doping mechanisms in polyaniline by optical, FTIR and Raman spectroscopy. Synthetic Metals. 1989;**29**:193

[37] Furukawa Y, Ueda F, Hydo Y, Harada I, Nakajima T, Kawagoe T. Vibrational spectra and structure of polyaniline. Macromolecules. 1988;**21**:1297

[38] Quillard S, Louarn G, Lefrant S, MacDiarmid AG. Vibrational analysis of polyaniline: A comparative study of leucoemeraldine, emeraldine, and pernigraniline bases. Physical Review B. 1994;**50**:12496

[39] Berrada K, Quillard S, Louarn G, Lefrant S. Polyanilines and substituted polyanilines: A comparative study of the Raman spectra of leucoemeraldine, emeraldine and pernigraniline. Synthetic Metals. 201;**1995**:69

[40] Louarn G, Lapkowski M, Quillard S, Pron A, Buisson JP, Lefrant S. Vibrational properties of polyaniline - Isotope effects. The Journal of Physical Chemistry. 1996;**100**:6998

Section 2

Assessment of Nanomaterial's Toxicity

Chapter 2

How to Assess Nanomaterial Toxicity? An Environmental and Human Health Approach

Matheus M. Roberto and Cintya A. Christofoletti

Abstract

Nanomaterials had been used because of the properties they exert in such scale (<100 nm), and they have been used in a wide variety of products like paints, electronics, fabrics, and also personal care products. Recent manuscripts available in the literature demonstrate the potential benefits of nanotechnology with these products. However, when released in the environment or when in contact with the human body, it is hard to track their final destination and their influence over the living beings. So, since nanomaterials were considered an important technology, a concern about its risks also started. Due to the variety of sizes, physicochemical properties, and uses, many researchers are aiming to assess the possible toxicity of this class of particles. Because of that, the chapter objective is to gather which assay, performed in vivo and in vitro, is the most frequently used and recommended to measure nanomaterial toxicity. Also, it is important to know which is the most suitable test to evaluate the toxicity over the environment, through direct effect and after biodegradation, and also related to human health. This chapter presents a concise review about the accepted methods to assess nanomaterial toxicity and also discuss about the need for regulamentation.

Keywords: toxicity assays, test organisms, in vitro models, in vivo assays, nanoscale assessment

1. Introduction

Emerging and promising nanotechnology represents a field of multidisciplinary knowledge responsible for development and application of materials, which measure less than 100 nm [1, 2]. The Royal Society and Royal Engineering Academy proposed this concept in 2004, which was associated to nanoscience as the branch responsible for studying the phenomenon of materials with atomic, molecular, and macromolecular scales, whose properties differ significantly from those with major scales [3, 4].

Nanoparticles can be generally described as ultrafine small material with 1–100 nm; however, several types of systems not limited only by small particles of certain material are included in this definition, as nanotubes, nanospheres, and nanocapsules [4, 5]. The properties exhibited by nanomaterials are unique and are being applied in many fields, from industrial to medicine [6, 7]. According to Arora et al. [8], the use of nanomaterials is increasing for commercial purposes as fillers, opacifiers, water filtration agents, cosmetic ingredients, semiconductors, electronic parts, and others. However, these same authors report that nanomaterials are being used in the medical area, mainly as agents for drug delivery, biosensors, and imaging contrast, i.e., human contact can happen both indirectly and directly, also being administered by ingestion or injection [8]. Once nanomaterials are used, environmental releasing turns dependent on the incorporation form of this product in each matrix, intrinsic material properties and also environmental conditions [9]. When there is human exposure or direct intake of nanomaterials, nanoparticles' physicochemical properties and its possible modifications can influence absorption, distribution, and organism metabolism. Besides the potential to accumulate in some organs, relevant rates of nanomaterials are excreted, being released to the environment [10]. About the nanomaterials presence in the environment, a detailed description regarding its sources and fates can be found in the review of Part [11].

Due to the new scale of some materials, new physicochemical interactions may occur bringing unexpected and also adverse effects because these elements generally become highly reactive [12]. Physicochemical properties observed in engineered nanomaterials are attributed to small size, chemical composition (purity, crystallinity, electronics characteristics, etc.), structural surface (reactivity, organic or inorganic coating, etc.), solubility, form, and agglomeration potential [8].

In view of the properties that the nanomaterials present, studies that evaluate the toxicity, their behavior in different environments, and the interactions with the biological system are of extreme importance. According to Dusinska [6], the safety assessment of nanomaterials is based on principles of risk assessment of "bulk" chemical substances. However, it is known that the behavior of these materials, both in the environment and in the cells, is different from such crude samples, and therefore the monitoring needs to be more specific. Catalán et al. [13] emphasize that the damaging potential of biodurable nanomaterials is not well demonstrated, and thus the classical toxicity evaluation trials must undergo adaptations.

2. Brief history of nanotoxicology

According to Maynard et al. [14], until the 1990s, many studies that focused on environmental epidemiology indicated a relationship between exposure to aerosols and increased mortality and morbidity of organisms. The relationships between particle size, chemical nature, and toxic effects were demonstrated, with the most pronounced effects observed in the lungs and heart due to exposure to smaller particles. These same authors argue that only in this decade has there been evidence that environmental particles with a diameter of less than 2.5 μm could cause deleterious health effects due to their reduced size [14]. Now it is known that engineered nanoparticles can perform these same activities [12].

Since the inception of this science, the studies and applications of nanoparticles have grown exponentially and, to the same extent, heightened concerns about environmental and health implications. In this context, the term nanotoxicology was formalized by a proposal of Donaldson et al. in 2004 [15] in an editorial in the journal *Occupational and Environmental Medicine* [5] and, since then, has been used to describe specifically the harmful effects of nanomaterials on environmental, animal, and human health. In 2005, nanotoxicology was consolidated as an area of expertise, with the launch of the journal *Nanotoxicology*, with the first article published by Oberdörster et al. in 2007 [16]. This article discusses the history of nanotoxicology as a science and presents some challenges to be faced by researchers.

Considering that nanoparticles have a greater potential to travel through the body than conventional-sized materials, researchers warn of the possibility of numerous interactions with biological fluids, cells, and tissues. Therefore, in vitro tests are recommended for an initial evaluation of the cytotoxicity and genotoxicity of nanomaterials, as well as for the identification and understanding of cellular mechanisms of toxicity [3]. In vivo methods are also used and, for both, some methods have already been developed by the Organization for Economic Co-operation and Development (OECD) and can be used for regulatory purposes.

3. Toxicological aspects of nanomaterials

According to Paschoalino et al. [3], the growing investment in nanoscience boosted the world market, as well as increased the use and consumption of products and processes aimed at this area. Despite this, it is true that research aimed at evaluating the toxicity of nanomaterials is still necessary, since the same properties that make nanomaterials so attractive may also be responsible for harmful effects on living organisms.

In this context, there is a recommendation for the analysis of physicochemical properties of nanomaterials in relation to human health and environmental safety (**Table 1**). In 2006, the OECD established a working party on manufactured nanomaterials to determine the appropriate methods for evaluating nanomaterials. According to the guidance manual developed, 26 physicochemical properties should be considered [6]. However, according to these same authors, only a few

Property	Relevance
Particle size distribution	Essential
Degree/state of agglomeration	Important
Particle shape	Important
Chemical composition/purity	Essential
Solubility	Essential (if applicable)
Physical properties	
Density	Matrix dependent
Crystallinity	Matrix dependent
Microstructure	Matrix dependent
Optical and electronic properties	Matrix dependent
Bulk powder properties (important for dosimetry/exposure)	Matrix dependent
Concentration (can be measured as mass, surface area, or number concentrations)	Essential
Surface properties	
Specific surface area/porosity	Essential
Surface chemistry/reactivity	Essential
Surface adsorbed species	Important
Surface charge/Zeta potential (especially in aqueous biological environment— may change according the environment)	Important
Surface hydrophobicity	Essential
Adapted from Powers et al. [17]	

Table 1.
Properties used for nanomaterial characterization regarding toxicity evaluation.

methods are available for the characterization of the toxicological properties of the nanomaterials, and the association of the effects with the physicochemical characterization is still a challenge.

Nanomaterials encompass a broad spectrum of materials with different physical, chemical, and biological properties. Thus, they do not constitute a homogeneous group and are usually defined by the type of core, which may be organic, such as fullerenes (carbon derivatives) and carbon nanotubes (single and/or multilayer), or inorganic, such as those of metal oxides (iron, zinc, titanium, etc.), metals (mainly gold and silver), and quantum dots [4].

According to Ju-Nam and Lead [4], some nanomaterials can have their surfaces manipulated in order to introduce specific functionalities for new applications. Thus, a vast field of possibilities opens up for materials with different properties and, therefore, also for infinite interactions with organisms and environment. However, the major challenge of nanotoxicology is to understand and prevent the risk of the use and/or exposure to nanomaterials that can cause toxicity by mechanisms not yet known or not yet explained by traditional toxicology [5].

Concern about the toxicity of nanomaterials lies primarily in production and commercialization on such a large scale as at present. Thus, the risk of these compounds reaching the different environmental compartments (atmosphere, water, and soil), becoming bioavailable, is very large [3, 17].

Since 2005, the European Commission Scientific Committee on Emerging and Newly Identified Health Risks (SCENIHR) has published reports on the impacts of nanoparticles on human health. The aforementioned committee focused its efforts on the evaluation of nanoparticles physically capable of entering the human body via inhalation, ingestion, and dermal absorption and reported that the size, shape, surface area, and chemical composition of the nanoparticle are closely associated with its toxicity. In addition, it has been explored how these characteristics affect bioavailability and interactions, as well as influence on exposure and dose. Therefore, the dose, the physicochemical properties, and the biokinetics are also important parameters to be evaluated when considering the toxicology of nanomaterials [14].

4. Deposition and interactions of nanomaterials with cells and the environment

In view of the numerous properties and characteristics of nanomaterials, products that are increasingly light, resistant, and often of lower cost are daily produced and marketed by the most different segments, such as electronic, medical, pharmaceutical, cosmetic, food, and agricultural [3]. In this context, when considering the ecotoxicology of particles whose components are nontoxic in the micro- or macrometric scales, studies that elucidate the routes of exposure and effects of nanomaterials on environmental compartments and different organisms are fundamental.

According to Laux et al. [7], the entry of nanomaterials into the environment occurs by the release of their components during use and by final disposal, so it is important to track and understand the kinetics and transformation of these materials in organisms and the environment. Knowledge of the influence of biopersistence on biokinetics and environmental fate is of utmost importance when determining the toxic potential.

When a nanomaterial comes in contact with the human body or the environment, it is difficult to track it again. In the environment, some nanomaterials such as metallic (e.g., Ag and Cu) and metal oxides (e.g., ZnO and Fe_2O_3) can dissolve rapidly, while others are more persistent (e.g., TiO_2, SiO_2, carbon nanotubes, and graphene) [18]. However, according to these authors, soluble nanomaterials present the best scenario of toxicity evaluation, since their behavior is generally similar to that presented by their ions. However, when cells internalize them, they can solubilize and release toxic metals through a mechanism known as "Trojan Horse" [18].

The aquatic ecosystem is the main route of exposure to a nanomaterial, since this type of environment is usually the final destination of nanocomposites introduced in natural systems [19]. After the aquatic environment, the atmosphere (troposphere), soil, and sediment follow an order of priority as routes of exposure. When present in aquatic environments, nanomaterials can be absorbed by cells, especially during filtration by aquatic organisms, directly interfering with their physiology and/or their ability to feed and breathe [3]. According to these same authors, other pathways of entry of nanomaterials into receptor organisms occur through cellular uptake, inhalation, or ingestion (**Figure 1**).

According to Bhaskar et al. [20] and Dusinska et al. [6], nanomaterials can enter the cells actively or passively, overcoming any protective barrier of the organism, including the blood-brain barrier. The capture mechanisms are related to the intrinsic physicochemical characteristics of the nanomaterial, as well as its route of exposure. Considering human health, generally the main route of exposure is inhalation, in which smaller particles reaching the alveoli and depending on their physicochemical properties cross the blood-air barrier of the lungs and reach the liver, heart, spleen, and kidneys [7]. There is a challenge when it comes to nanoparticles developed to cross human body barriers, such as those applied in medicine. There are materials that are developed to pass through barriers, not to enter cells, while there are others that are designed to act within them [10].

The first contact of the cell with any extracellular material occurs through the lipid (e.g., phospholipid) and protein components (e.g., membrane receptors) present in the cell membrane. For Paschoalino et al. [3], the nanomaterials present greater permeability to the skin, mucous membranes, and cell membranes due to their diminutive size.

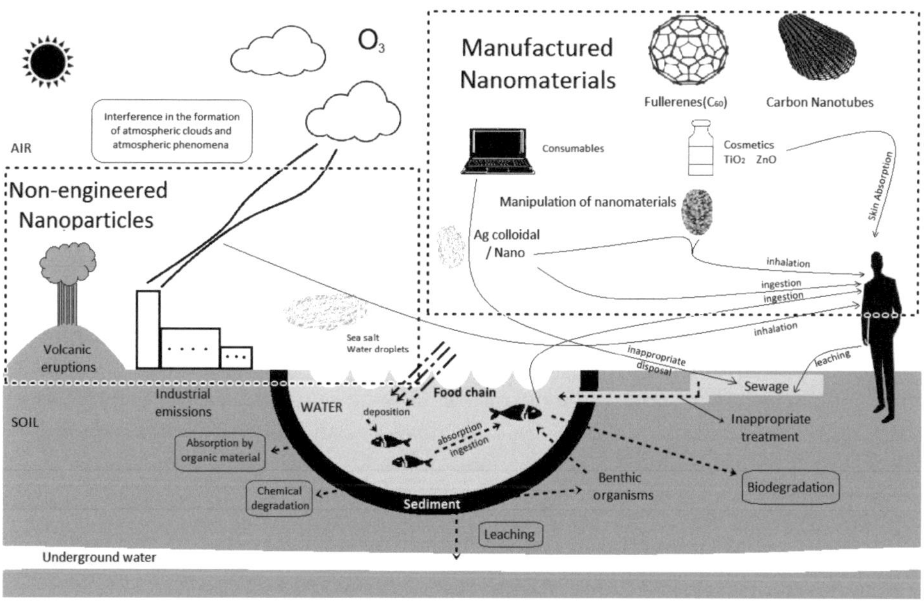

Figure 1.
Main sources, routes of exposure, and possible interactions between nanoparticles with the environment and organisms (adapted from Paschoalino et al. [3]).

Conner and Schmid [21] stated that most nanomaterials are actively incorporated by cells through endocytosis. This is one of the most important mechanisms of cellular communication with the external environment, since it involves the transmembrane and bidirectional flow of vesicles, through the movement of extracellular content internalization [1]. According to Radaic et al. [1], the shape, size, characteristics (such as porosity) of the surface, surface charge, and composition of nanoparticles directly influence endocytosis. According to Drasler et al. [22], cell size, proliferation rate, and surface receptor growth and expression characteristics are the major factors involved in the entry of nanomaterials into cells. Generally, endocytic pathways are essential in this process, where large particles or agglomerates of nanomaterials are obtained by phagocytosis (diameter greater than 250 nm), whereas smaller particles (diameter smaller than 150 nm) are obtained by pinocytosis, specific or not. Valsami-Jones and Lynch [18] have beautifully illustrated the possible mechanisms of nanomaterial uptake by cells.

Collins et al. [23] have described that besides penetrating cells, many nanomaterials are able to cross nuclear membranes and gain access to chromatin at any stage of the cell cycle. Thus, in addition to direct damage to DNA, nanoparticles can induce the formation of reactive oxygen species (ROS), such as hydroxyl radicals ($^{\cdot}OH$), causing oxidative stress (redox imbalance) and serious damage to the cell. Oxidative stress can be the result of the simple cellular response to the presence of the nanomaterial or a secondary effect of the inflammation generated by them [23]. It is also known that the dissolution of certain nanomaterials may be able to release toxic ions and/or other components, which may induce toxicity [22].

ROS are highly reactive molecules that disrupt intracellular medium homeostasis, since they interact with cellular macromolecules such as DNA, proteins, and lipids [24]. Singh et al. [24], Louro et al. [25], and Radaic et al. [1] stated that nanomaterials can induce genotoxic damage mediated by oxidative stress and through their interaction with cellular constituents, including mitochondria and NADPH oxidases bound to the cell membrane; by the depletion of antioxidants; or by the release of the metallic ions present in the constitution of many nanomaterials, which can promote the conversion of cellular oxygen metabolites into ROS. When considering the DNA molecule, the major damage induced by ROS is single-strand breaks, double-strand breaks, base modifications (such as the formation of 8-hydroxydeoxyguanosine adducts), and DNA cross-links. According to Singh et al. [24], all of the aforementioned damages have the potential to initiate and promote carcinogenesis. Once the DNA molecule has been damaged, several cellular processes can be triggered, such as cell cycle arrest, apoptosis, or DNA repair [24].

Apart from the oxidative stress-related lesions, other genotoxic effects may contribute significantly to the promotion of genetic instability, as nanomaterials can cross the pores of the nuclear envelope and interact directly with the genome of the cell and/or with nuclear proteins [25]. Under these conditions, Louro et al. [25] reported that some nanomaterials induce the formation of intranuclear protein aggregates, inhibiting the processes of cell replication, transcription, and proliferation. When the nanomaterials are not able to cross the nuclear envelope, there is still the possibility of interaction with the DNA molecule and nuclear proteins during the mitotic process, which can cause aneuploidies [24, 25].

Depending on the organism exposed to the nanomaterials, different interactions can be evidenced. Bielmeyer-Fraser et al. [26] demonstrated that nanoparticles of ZnO, AgO, and CuO were able to induce toxicity to algae bioindicator in a similar way to the respective solubilized metals. However, these researchers noted the metals accumulated in different regions of the cells, and the nanoparticles were retained mainly in the cell wall, while the metals were observed mainly in the organelles, as fragments of the endoplasmic reticulum. However, in both ways, the authors point out that the accumulated metal could be transferred by the trophic chain and carried to other organisms.

When a nanomaterial enters a living organism, several components can adhere to its surface, drastically modifying its interaction with cellular structures. Proteins are molecules that can adhere to the nanomaterial and form a type of coating, called biomolecular corona [27]. This corona may alter the ability of a nanomaterial to cross physiological barriers, influencing its toxic potential [7].

According to Drasler et al. [22], in vitro assays performed with cell culture can indicate the biological fate of nanomaterials at the cellular and multicellular level, even in an excluding mode according to the cell type (i.e., to determine which cell type is actually affected by a certain type of nanomaterial).

5. Methods for evaluating the toxicity of nanomaterials

Within nanotoxicology there is an impasse on how best to assess the possible adverse effects of nanomaterials, both for human health and environmental monitoring. Toxicity tests may be performed employing live (in vivo) organisms, such as microcrustaceans, fishes, rodents, and other animals and/or cell cultures (in vitro). Several standardized toxicological tests are available to measure the biological response of an organism to a chemical. However, there is no standardization for the evaluation of the toxicity of nanomaterials, which hampers the comparison of results and the consensus about their toxicity. Most of the studies performed so far are adaptations of the standard procedures used for other substances [3]. Although some minimal combinations of assays have been proposed, Drasler et al. [22] have described that there is no standard evaluation protocol due to the wide range of physicochemical properties that nanomaterials can present.

Animal tests are more predictive for human effects but have limitations, mainly because of physiological and biochemical differences between the species. In addition, there is a growing public and legal demand that ethically supports the substitution of animal testing for alternatives not based on in vivo testing. New concepts of experimentation have been based on strategies with primary culture of human cells and permanent cultures of well-established cell lines, since they present efficient, cheap, and reliable results [22].

Understanding the demand for orientation and applicability, this chapter will address some of the main evaluation methods, developed both in vivo and in vitro, to better characterize the toxicity of nanomaterials.

5.1 In vivo methods

In vitro evaluations have increased considerably, but in vivo validation still is necessary to understand and interpret its results. Furthermore, animal experimentation was also part of the NanoTEST project, whose purpose was to understand the effects on the physiology of organisms tested. Currently, the OECD presents some test guidelines on which biomarkers should be used for each test organism [10].

In general, there are more researches on human toxicity, using rodent models, whereas few in vivo studies addressing the ecotoxicity of nanomaterials are available. Furthermore, most of those found in the literature consider the impact of nanomaterials on aquatic organisms, since the continental and marine waters end up being the main receiving compartment. Some scarce trials address the toxicity of nanomaterials in soil and in atmosphere, commonly as suspended particles. In general, bacteria (e.g., *Aliivibrio fischeri*), algae (e.g., *Raphidocelis subcapitata*), nematodes (e.g., *Caenorhabditis elegans*), microcrustaceans (e.g., *Daphnia magna*, *D. pulex*, *Ceriodaphnia dubia*), mollusks (e.g., *Lymnaea stagnalis*), fish (e.g., *Danio rerio*), and rodents (Wistar rat and mice) are the most used test organisms for the evaluation of acute toxicity (**Table 2**).

5.2 In vitro methods

According to Drasler et al. [22], assays can be performed with primary cultures or eternal cell lines. According to these authors, cell lines are preferably chosen because they present great homogeneity and stability, which favors reliability in the results, especially in initial tests. For more specific tests, these same researchers recommend the use of 3D co-cultures, to better understand the mechanisms of action of nanomaterials on tissues.

For the nanomaterial toxicity evaluation, the use of epithelial cell lines (skin, gastrointestinal tract, or lung) is usually indicated as these cells present characteristics of real barriers against harmful agents and are therefore the first to suffer the influence of these compounds [37]. However, it is important to note that some strains may not be responsive to the effects of nanomaterials and, in this case, primary cultures may be more indicated [22].

Aiming at the reproducibility of in vitro assays with culture of cell lines, it is necessary to record details that are generally missing from the publications. The origin of the cells, the number of the passage, the detailed method of cell culture, the brand of plastics, and reagents used during the cultivation/exposure, besides the description of the morphology, growth, and cell differentiation, before and after the test, are the information that should be included in the results' publication [22]. Among the in vitro assays, those performed with mammalian cells are considered to be more important than those performed with other cell types [13].

For the in vitro comet assay with mammalian cell culture, Collins et al. [23] make some recommendations: (1) use non-cytotoxic concentrations (less than 20% of cell viability loss; if the nanomaterial is not cytotoxic, concentrations below 100–150 μg/mL are recommended); (2) choose the cell type according to the exposure scenario (based on exposure route and target organ); (3) determine both short (2–3 h) and long (24 h) tests to obtain a better understanding of the mode of action of the nanomaterial; and (4) determine if the genotoxic damage evidenced is a result of the direct effect with the DNA or due to the oxidation of the DNA. According to Drasler et al. [22], the exposure period is one of the main factors related to contradictory toxicity results for identical nanomaterials, as this involves transformations and the aging of their components.

In vitro assays can cover specific endpoints, such as dermal absorption, skin and eye irritation, endocrine disruption, and genotoxicity, among others. Among the tests, most nanomaterial evaluation protocols align the main routes of exposure, being dermal, oral, and inhalation [22].

According to Catalán et al. [13], the relevance and limitations of genotoxicity/mutagenicity assays should be taken into account when choosing the most appropriate monitoring method. According to these authors, the tests considered in the evaluation should be based on three categories, following the importance order: (1) gene mutation, (2) chromosomal damage, and (3) DNA damage. DNA damage is considered a mild effect because of the possibility of repair, while chromosomal damage and gene mutation are considered to be severe effects because they are irreparable changes.

Regarding the mutagenic potential of nanomaterials to humans, the effects observed in vivo should be considered more relevant than those observed in vitro,

Nanomaterial	Mean diameter of the particles (nm)	Test organism	Main results	References
Ag	13–17 nm	*Lymnaea stagnalis* (Mollusca)	Growth alteration and bioaccumulation	Croteau et al. [28]
ZnO TiO_2	15–30 nm	*Skeletonema marinoi* (Diatom—Skeletomataceae), *Thalassiosira pseudonana* (Diatom—Thalassiosiraceae), *Dunaliella tertiolecta* (Algae—Dunaliellaceae), *Isochrysis galbana* (Algae—Isochrysidaceae)	Only nanoparticles of ZnO have decreased growth rate of diatom and algae population	Miller et al. [29]
Graphene family nanoparticles	—	*Caenorhabditis elegans* (Nematoda)	Decreased reproduction rates	Chatterjee et al. [30]
ZnO CuO AgO	ZnO: 20–30 nm CuO: 20–100 nm AgO: 20–70 nm	*Thalassiosira weissflogii* (Diatom—Thalassiosiraceae)	Decreased diatom population growth in similar way to respective dissolved metals. Bioaccumulation of nanoparticles in cell wall and possible transfer through trophic chain	Bielmeyer-Fraser et al. [26]
ZnO Al_2O_3 TiO_2	—	*Danio rerio* (Chordata)	Metal oxide nanoparticles induced different toxic effects in zebrafish development according to each metal. ZnO delayed larvae and embryo development and also induced serious ulceration in larvae	Zhu et al. [31]
TiO_2	~43 nm	*Pimephales promelas* (Chordata)	Fish immunotoxicity and gene expression alteration	Jovanović et al. [32]
TiO_2	5, 10, and 32 nm	*Xenopus laevis* (Chordata)	Significantly affected tadpole growth. The highest concentration caused mortality, suppressed tadpole body length, and delayed animal development	Zhang et al. [33]

Nanomaterial	Mean diameter of the particles (nm)	Test organism	Main results	References
TiO_2	—	*Daphnia similis* (Crustacea)	The highest concentration (100 mg L^{-1}) did not induce toxic effects under experimental conditions. A mixture of TiO_2 forms induced toxic effects by ROS generation when exposed to UVA light	Marcone et al. [34]
TiO_2 ZnO CuO	TiO_2: 25–70 nm ZnO: 50–70 nm CuO: 30 nm	*Vibrio fischeri* (*Gammaproteobacteria*), *Daphnia magna* (Crustacea), *Thamnocephalus platyurus* (Crustacea)	Suspensions of nano- and bulk TiO_2 were not toxic. A nano-ZnO formulation was very toxic to *V. fischeri*, *D. magna*, and *T. platyurus*. Cu compound also showed toxicity; however, for Daphnia magna were less bioavailable than for bacteria	Heinlaan et al. [35]
Metallic nanoparticles of Ag, Cu, Al, Co, Ni and TiO_2	Ag (20–30 nm), Cu (15–45 nm), Al (51 nm), Co (10–20 nm), Ni (5–20 nm), and TiO_2 (30 nm)	*Raphidocelis subcapitata* (Algae—Selenastraceae), *Ceriodaphnia dubia* (Crustacea), *Daphnia pulex* (Crustacea), *Danio rerio* (Chordata)	Nanometals caused acute toxicity in multiple aquatic organisms, but the effect was different according to the metal particle and the species used. Since *R. subcapitata*, *C. dubia*, and *D. pulex* were susceptible to nanometals, trophic chain could be compromised	Griffitt et al. [19]
Ag ZnO TiO_2 CeO_2 Cu	Ag (15 nm) ZnO (34–42 nm) TiO_2 (10–23 nm) CeO_2 (10–33 nm) Cu (76 nm)	*Raphidocelis subcapitata* (Algae—Selenastraceae), *Daphnia magna* (Crustacea), *Danio rerio* (Chordata)	Ag and Cu nanoparticles affected all organisms; ZnO was toxic to algae and daphnids; TiO_2 and CeO_2 were toxic only to algae	Hund-Rinke et al. [36]

Table 2.
Experimental conditions and obtained results through in vivo tests.

since the first allow the detection of inflammation and, therefore, secondary genotoxic effects [13]. Although more predictive for human effects, animal tests still have limitations, mainly because of the physiological and biochemical differences between species. Also, there is a trend to substitute animal testing for suitable alternatives that do not promote pain and suffering [22].

Among the tests recognized by the scientific community, those with certified guidelines for nanomaterial assessment have greater "weight" than others that have not been validated in the determination of genotoxicity/mutagenicity [13]. Although they cannot be used to determine mutagenicity, the remaining assays can be used to demonstrate the genotoxic potential of nanomaterials.

There are several recommended tests to assess nanomaterials, especially those described by the OECD. In accordance with the OECD guidelines [38], in order to select a test and evaluate the genotoxicity of a nanoform, exposure, absorption, solubility, metabolites, and other derivatives should be considered, as well as possible side effects (e.g., generation of ROS).

Comparing the genotoxicity tests for chemical substances, the comet assay and the micronucleus test are also the most indicated and used by the researchers [13]. The comet assay (single cell gel electrophoresis) is a common method of DNA damage evaluation, which can be performed with very diverse cell types. Briefly, a suspension of individualized cells is mixed with agarose and placed on a pregelatinized slide. Then, cell lysis on Triton X-100 removes membranes and soluble cellular components, while NaCl removes the histones from the DNA, promoting a superadhesion of this material to a matrix, forming a structure known as a nucleoid. When there are breaks in DNA strands (single or double), the fragments tend to move toward the anode during electrophoresis. When there is damage and it is observed by fluorescence microscopy, a comet-like image is noted. The percentage of DNA in the tail is proportional to the frequency of breaks, that is, the damage inferred to the genetic material [23].

As described by Collins et al. [23], several nanomaterials (e.g., TiO_2, ZnO, Au, Ag, Co_3O_4, Fe_3O_4, SiO_2, ZrO_2, and others) have already been evaluated by variations of the comet assay with specific endonucleases for some lesions, which increase the power of this tool. Among these enzymes, formamidopyrimidine DNA glycosylase (FPG) recognizes lesions of the 8-oxo-7,8-dihydroguanine (8-oxoG) and formamidopyrimidine type (open-ring purines) and is therefore widely used to estimate oxidative damages to DNA caused by nanomaterials [23].

However, other famous trials are not recommended, such as the Ames test [13]. Catalán et al. [13] discourage the use of this test to evaluate nanomaterials, since some compounds are unable to cross the bacterial wall, while others have bactericidal effect.

6. Final considerations

Undoubtedly, nanoscience and nanotechnology offer the prospect of great advances to the most different sectors of industry and medicine. However, as any area of technology that makes intensive use of new materials/structures, it brings some risks to the health of organisms and the environment. Generally, toxicological studies involving nanomaterials are still scarce, with results often controversial when compared to each other, mainly due to incipient standardization. In this context, the combination of in vitro and in vivo methods in a battery of tests is still the best way to assess the toxicity of nanomaterials [22, 23].

One of the major concerns is the choice of dose/concentration range of nanomaterials to be tested. The inclusion of excessively high doses/concentrations may generate false positives, while excessively low doses may prevent detection or may underestimate the genotoxic potential [23]. Drasler et al. [22] provide all guidelines to be considered in evaluating the toxicity of nanomaterials by cell culture, but in vivo evaluation must not be overlooked. Paschoalino et al. [3] state that the environmental risk analysis of nanomaterials depends mainly on the regulatory structure, which involves the generation of protocols, which must be based on a multidisciplinary interaction, in order to obtain a more risk assessment possible.

As demonstrated by Valsami-Jones and Lynch [18], harmonization of methods and approaches could benefit this young science, as there is still no consensus on basic assessment protocols. Current protocols involve specific techniques and methods to collect and analyze data sufficient to quantitatively describe the release, destination, transport, transformation, exposure, and toxicity of chemicals. Furthermore, in order to be more precise about the toxicity and mechanism of action of nanomaterials on living organisms, the physicochemical characteristics must be sufficiently detailed. So far, a great effort has been made by the OECD to try to standardize test methods that can correctly evidence the risk of nanomaterials. There are a number of internationally accepted test guidelines that are used for toxicity assessment involving trials with organisms for aquatic, soil, and sediment monitoring. Since 2013, experts from all over the world hold strategic meetings to determine what directions the OECD should take regarding the assessment of nanomaterials, as explored in the Petersen et al. [39].

Also, there is a lot of potential in computational models to help elucidate the possible effects of nanomaterials on humans and the environment. Currently, the quantitative structure-activity relationship (QSAR) model seems to be quite adequate because it can relate the structural, physical, and chemical characteristics to the behavior that some nanomaterial can present. To date, the combination of field, laboratory, and computational work still is the most promising technique to ensure reliable responses to the issues involved with nanomaterial toxicity.

Conflict of interest

The authors declare no conflict of interest.

Author details

Matheus M. Roberto* and Cintya A. Christofoletti
Hermínio Ometto Foundation's University Center, Araras, Brazil

*Address all correspondence to: mmr@fho.edu.br

References

[1] Radaic A, Pugliese GO, Campese GC, Pessine FBT, Bispo de Jesus M. Como estudar interações entre nanopartículas e sistemas biológicos. Química Nova. 2016;**39**(10):1236-1244

[2] Dusinska M, Rundén-Pran E, Schnekenburger J, Kanno J. Toxicity tests: In vitro and in vivo. In: Adverse Effects of Engineered Nanomaterials. 2nd ed. Cambridge: Academic Press; 2017. pp. 51-82

[3] Paschoalino MP, Marcone GPS, Jardim WF. Os nanomateriais e a questão ambiental. Quimica Nova. 2010;**33**(2):421-430

[4] Ju-Nam Y, Lead JR. Manufactured nanoparticles: An overview of their chemistry, interactions and potential environmental implications. Science of the Total Environment. 2008;**400**(1-3):396-414

[5] Consiglieri VO. Biossegurança em nanotecnologia. In: Hirata MH, Hirata RDC, Mancini Filho J, editors. Manual de Biossegurança. 2nd ed. Barueri: Manole; 2012. pp. 331-339

[6] Dusinska M, Tulinska J, Yamani El N, Kuricova M, Liskova A, Rollerova E, et al. Immunotoxicity, genotoxicity and epigenetic toxicity of nanomaterials: New strategies for toxicity testing? Food and Chemical Toxicology. 2017;**109**(Part 1):797-811

[7] Laux P, Riebeling C, Booth AM, Brain JD, Brunner J, Cerrillo C, et al. Biokinetics of nanomaterials: The role of biopersistence. NanoImpact. 2017;**6**:69-80

[8] Arora S, Rajwade JM, Paknikar KM. Nanotoxicology and in vitro studies: The need of the hour. Toxicology and Applied Pharmacology. 2012;**258**(2):151-165

[9] Suzuki S, Part F, Matsufuji Y, Huber-Humer M. Modeling the fate and end-of-life phase of engineered nanomaterials in the Japanese construction sector. Waste Management. 2018;**72**:389-398

[10] Juillerat-Jeanneret L, Dusinska M, Fjellsbø LM, Collins AR, Handy RD, Riediker M, et al. Biological impact assessment of nanomaterial used in nanomedicine. Introduction to the NanoTEST project. Nanotoxicology. 2015;**9**(Suppl. 1):5-12

[11] Part F, Berge N, Baran P, Stringfellow A, Sun W, Bartelt-Hunt S, et al. A review of the fate of engineered nanomaterials in municipal solid waste streams. Waste Management. 2018;**75**:427-449

[12] Oberdörster G, Oberdörster E, Oberdörster J. Nanotoxicology: An emerging discipline evolving from studies of ultrafine particles. Environmental Health Perspectives. 2005;**113**(7):823-839

[13] Catalán J, Stockmann-Juvala H, Norppa H. A theoretical approach for a weighted assessment of the mutagenic potential of nanomaterials. Nanotoxicology. 2017;**11**(8):964-977

[14] Maynard AD, Warheit DB, Philbert MA. The new toxicology of sophisticated materials: Nanotoxicology and beyond. Toxicological Sciences. 2011;**120**(Supplement 1):S109-S129

[15] Donaldson K, Stone V, Tran CL, Kreyling W, Borm PJA. Nanotoxicology. Occupational and Environmental Medicine. 2004;**61**(9):727-728

[16] Oberdörster G, Stone V, Donaldson K. Toxicology of nanoparticles: A historical perspective. Nanotoxicology. 2007;**1**(1):2-25

[17] Powers KW, Palazuelos M, Brown SC, Roberts SM. Characterization of nanomaterials for toxicological evaluation. In: Sahu SC, Casciano DA, editors. Nanotoxicity: From In Vivo And In Vitro Models To Health Risks. New Jersey: John Wiley & Sons. 2009. pp. 1-27

[18] Valsami-Jones E, Lynch I. NANOSAFETY. How safe are nanomaterials? Science. 2015;**350**(6259):388-389

[19] Griffitt RJ, Luo J, Gao J, Bonzongo J-C, Barber DS. Effects of particle composition and species on toxicity of metallic nanomaterials in aquatic organisms. Environmental Toxicology and Chemistry. 2008;**27**(9):1972-1978

[20] Bhaskar S, Tian F, Stoeger T, Kreyling W, la Fuente de JM, Grazú V, et al. Multifunctional nanocarriers for diagnostics, drug delivery and targeted treatment across blood-brain barrier: Perspectives on tracking and neuroimaging. Particle and Fibre Toxicology. 2010;**7**(1):3

[21] Conner SD, Schmid SL. Regulated portals of entry into the cell. Nature. 2003;**422**(6927):37-44

[22] Drasler B, Sayre P, Steinhäuser KG, Petri-Fink A, Rothen-Rutishauser B. In vitro approaches to assess the hazard of nanomaterials. NanoImpact. 2017;**8**:99-116

[23] Collins A, Yamani El N, Dusinska M. Sensitive detection of DNA oxidation damage induced by nanomaterials. Free Radical Biology and Medicine. 2017;**107**:69-76

[24] Singh N, Manshian B, Jenkins GJS, Griffiths SM, Williams PM, Maffeis TGG, et al. NanoGenotoxicology: The DNA damaging potential of engineered nanomaterials. Biomaterials. 2009;**30**(23-24):3891-3914

[25] Louro H, Borges T, Silva MJ. Nanomateriais manufaturados: Novos desafios para a saúde pública. Revista Portuguesa de Saúde Pública. 2013;**31**(2):188-200

[26] Bielmyer-Fraser GK, Jarvis TA, Lenihan HS, Miller RJ. Cellular partitioning of nanoparticulate versus dissolved metals in marine phytoplankton. Environmental Science & Technology. 2014;**48**(22):13443-13450

[27] Emer M, Cardoso MB. Biomolecular corona formation: Nature and bactericidal impact on surface-modified silica nanoparticles. Journal of Materials Chemistry B. 2017;**5**(40):8052-8059

[28] Croteau M-N, Misra SK, Luoma SN, Valsami-Jones E. Silver bioaccumulation dynamics in a freshwater invertebrate after aqueous and dietary exposures to nanosized and ionic Ag. Environmental Science & Technology. 2011;**45**(15):6600-6607

[29] Miller RJ, Lenihan HS, Muller EB, Tseng N, Hanna SK, Keller AA. Impacts of metal oxide nanoparticles on marine phytoplankton. Environmental Science & Technology. 2010;**44**(19):7329-7334

[30] Chatterjee DK, Fong LS, Zhang Y. Nanoparticles in photodynamic therapy: An emerging paradigm. Advanced Drug Delivery Reviews. 2008;**60**(15):1627-1637

[31] Zhu X, Zhu L, Duan Z, Qi R, Li Y, Lang Y. Comparative toxicity of several metal oxide nanoparticle aqueous suspensions to Zebrafish (*Danio rerio*) early developmental stage. Journal of Environmental Science and Health. Part A, Toxic/Hazardous Substances & Environmental Engineering. 2008;**43**(3):278-284

[32] Jovanović B, Anastasova L, Rowe EW, Zhang Y, Clapp AR, Palić D. Effects of nanosized titanium dioxide on innate immune system of fathead minnow

(*Pimephales promelas* Rafinesque, 1820). Ecotoxicology and Environmental Safety. 2011;**74**(4):675-683

[33] Zhang J, Wages M, Cox SB, Maul JD, Li Y, Barnes M, et al. Effect of titanium dioxide nanomaterials and ultraviolet light coexposure on African clawed frogs (*Xenopus laevis*). Environmental Toxicology and Chemistry. 2012;**31**(1):176-183

[34] Marcone GPS, Oliveira ÁC, Almeida G, Umbuzeiro GA, Jardim WF. Ecotoxicity of TiO_2 to Daphnia similis under irradiation. Journal of Hazardous Materials. 2012;**211-212**:436-442

[35] Heinlaan M, Ivask A, Blinova I, Dubourguier H-C, Kahru A. Toxicity of nanosized and bulk ZnO, CuO and TiO_2 to bacteria *Vibrio fischeri* and crustaceans Daphnia magna and *Thamnocephalus platyurus*. Chemosphere. 2008;**71**(7):1308-1316

[36] Hund-Rinke K, Schlich K, Kühnel D, Hellack B, Kaminski H, Nickel C. Grouping concept for metal and metal oxide nanomaterials with regard to their ecotoxicological effects on algae, daphnids and fish embryos. NanoImpact. 2018;**9**:52-60

[37] Rothen-Rutishauser B, Clift MJD, Jud C, Fink A, Wick P. Human epithelial cells in vitro – Are they an advantageous tool to help understand the nanomaterial-biological barrier interaction? EURO-NanoTox-Letters. 2015;**4**(1):1-19

[38] OECD. Draft guidance document on the use of OECD test guidelines on genotoxicity testing for manufactured nanomaterials. Organization for Economic Co-operation and Development. 2014. pp. 1-3. http://www.oecd.org/env/ehs/testing/Draft_GD_nano_and_genotox_rev2.pdf

[39] Petersen EJ, Diamond SA, Kennedy AJ, Goss GG, Ho K, Lead J, et al. Adapting OECD aquatic toxicity tests for use with manufactured nanomaterials: Key issues and consensus recommendations. Environmental Science & Technology. 2015;**49**(16):9532-9547

Chapter 3

Risk Assessment and Health, Safety, and Environmental Management of Carbon Nanomaterials

Guilherme Lenz e Silva, Camila Viana, Danieli Domingues and Fernanda Vieira

Abstract

The management of health, safety, and environmental (HSE) aspects during production, manipulation, storage, incorporation, and disposal of carbon nanomaterials is the key factor for the development of a safe-by-design work based on nanotechnology. The almost endless possibility of functionalization, chemical interaction, and addition of nanomaterials into new products implies a new management approach of HSE. Low amount of reliable toxicity and ecotoxicity data of nanomaterials and nanomaterial composites is available. As complete exposure/release assessments are a challenging task, recommendation for control measurements is still based on the precautionary point of view. There is an incomplete understanding of environmental fate- and time-related exposure, and of consumer- and worker-related risks and hazards. Control banding and risk evaluation matrix tools can be used to mitigate labor and environment impacts of carbon nanomaterials. This chapter presents new tools and methodologies for exposure assessment and risk evaluation of hazards used on HSE management system of carbon nanomaterials.

Keywords: risk assessment, risk analysis, risk evaluation, nanocarbons, nanotechnology, HSE

1. Introduction

Looking into the last 20 years—since the classical publications from the Royal Society, ETC Group, OECD/Allianz, European Commission, Swiss Reinsurance, and studies and think tanks related to risks and uncertainties of nanotechnology [1–7]—without any doubt, we are now in a better position concerning risks (assessment, analysis, evaluation, management, communication, policies, monitoring, and treatment) and benefits of nanotechnologies. However, new nanomaterials and new nanoproducts are achieving several markets, and the global changes on the business environment—including, fast, small and short-term life enterprises, start-ups and spin-offs based on new technologies—underline the need of special attention from the society regarding safety and environmental issues. Sometimes, in the nanotechnology business endeavor, low capital and poor-trained people without enough knowledge and guidance about safety, health and environmental regulations, standards and protocols could exposure, unnecessarily, workers, consumers and the environment.

From public and private investments, thousands of publications, research efforts, and scientific projects (money, time, and minds) have been done and much more should be done to follow the rapid technological and scientific development of nanoscience and its applications.

The knowledge of bio and physicochemical interactions between nanomaterials and living organisms, the understanding of life cycle analysis (LCA) of nanomaterials on environment, and the time-related effects of chronic dose-response toxicity are key points that should be taken in account during the debate and to improve the safety approach of nanotechnological development.

Risk assessment and management of health, safety, and environmental (HSE) nanotechnology are very important aspects to mitigate risks and improve the benefits and transform opportunities into technological development on medical applications (cancer treatment, tissue engineering, diagnosis of diseases, DNA manipulation, etc.), flexible and communication devices, portable energy, food conservation, agriculture productivity and pest control, and countless new applications and uses.

Carbon is one of the most import elements on Earth. Carbon nanomaterials are changing the technology of XXI century, and its fundamental electronic and hybridization states transform it into one of the most remarkable chemical elements with applications from regenerative medicine to rocket capsules and flexible electronic devices.

2. Carbon nanomaterials

Carbon hybridization states with sp^3 and sp^2 bonding system ally with the allotropy of carbon materials that produces soft, hard, light, and dense materials.

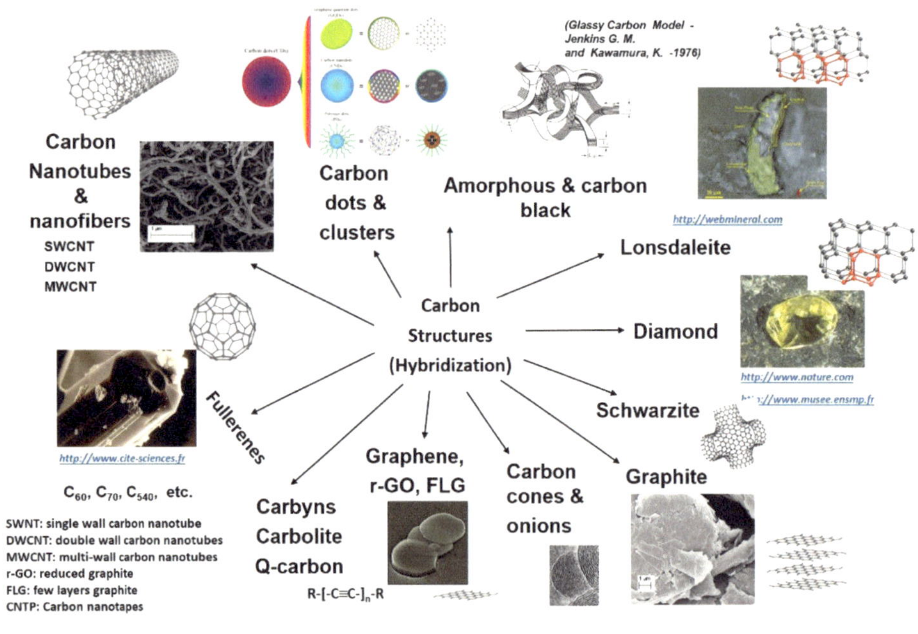

Figure 1.
Carbon allotropes family (adapted from Ref. [12]).

The carbon use and applications came from charcoal for heating, passed by nuclear reactors to carbon-carbon composites used on new aircrafts or space capsules. However, after the discovery of fullerenes from Curt et al. [9, 10] and discovery of graphene by Geim and Novoselov [11], the research of carbon nanomaterials has been expanding in an unimaginable way [8–11]. **Figure 1** shows the carbon allotropes with special attention on nanoforms.

The hybridization states of carbons allow then to assume different geometries in the space from 0D dimension (fullerenes, carbon dots) to 1D fiber type materials with high aspect ratio, passing by 2D flat material (graphene) to high complex 3D material such graphite, diamond, or schwarzite materials.

3. Health, safety, and environmental (HSE) management of nanomaterials

Why we should think differently about health, safety, and environment when dealing with nanomaterials or nanoobjects? Nanotechnology has defined areas of science that take place at nanoscale (1–100 nm) [13, 14]. Many specific and fundamental aspects are related to HSE, including surface energy, functionability, and size. **Figure 2** shows the main nanoparticle properties and how they are related regarding toxicological and risk assessment of nanomaterials. Those properties and the relation between them can be considered main aspects that should be considered to asses specific nanoparticle's risks.

Size (distribution), shape, surface area, chemical composition/crystalline structure, and solubility are very important properties regarding the risk assessment of carbon nanomaterials [15].

These properties can be assessed using different analytical tools. **Table 1** shows the nanomaterial properties, the main characterization techniques, and also the degree of importance on the knowledge of such property for a risk assessment point of view.

It is noticed that the concentration of nanomaterial could be described using different metrics such as mass/volume or number of particulates (or fibers)/volume.

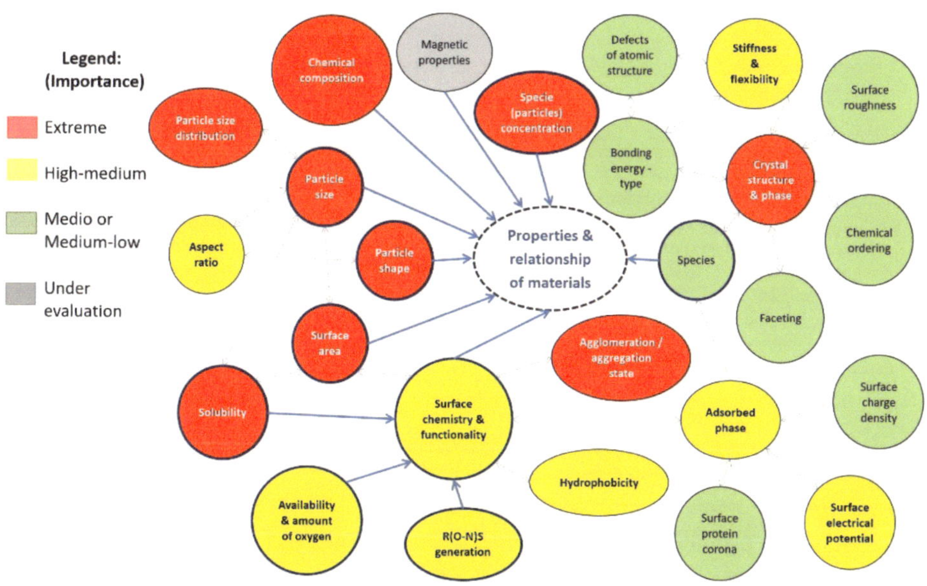

Figure 2.
Major relationships between properties and risk/toxicological interactions (adapted from Ref. [15]).

Nanomaterial properties	Importance	Main analytical characterization techniques
Chemical composition, structure and crystalline phase	Extreme	XRD, XRF, ICP-AES, AAS, HPLC, SIMS, MALD-TOF, EDS
Specific surface area (density, porosity, and reactivity: flammability and explosivity)	Extreme	Multimolecular gas adsorption (BET/adsorption isotherm method), gas (He) pycnometer, mercury intrusion porosimetry (MIP)
Shape dimensionality/morphology (0D, 1D, 2D, or 3D), number of layers	Extreme	TEM, XRD, SEM, Raman
Stiffness (rigid and flexibility, aspect ratio)	High-medium	Microscopy (SEM, TEM, ...), XRD
Solubility (dissolution and toxicant species) release	Extreme	Tables of solubility, Hildebrand and Hasen parameters
Surface species: R(O-N)S: reactive (oxygen and nitrogen) species, C/O atomic ration, etc.	High-medium	XPS, FTIR, NBT test, ESR, EPR
Surface chemistry (functionality, hydrophobicity and hydrophilicity, and adsorbed compounds and groups)	High-medium	Oxygen singlet analysis, XPS, FTIR, pH, contact angle microscopy, AES, EELS, DTA
Size and size distribution (aerodynamic, hydrodynamic)	Extreme	SEM, DLS, FCS, XRD
Solid bonding energy and type	Medium-low	DSC-TGA, simulation (first principles)
Surface charge density	Medium	SPM, PCM, SPR
Surface speciation	Medium-low	XPS, FTIR, NMR
Agglomeration and aggregation state	High-medium	XRD, SEM
Surface nanoparticle-protein corona	Medium	Simulation, electrophoresis
Bio-(persistence, accumulation, degradation and digestibility), endotoxin content	High-medium	Specific corrosion, digestibility, chemical analysis, LAL, TET, TNF, etc.
Structural atomic defects, ordering, and faceting	Medium-low	Microscopy (TEM, SEM), XRD, Raman
Quantum, electric and magnetic properties	Under evaluation	Induced magnetization, Currie balance
Concentration (mass, number of particles/fibers): property of area: environment, workplace, office, etc.	Extreme	OPC, CPC, light-scattering photometer, gravimetric analysis, etc.

Legend: XPS: X-ray photoelectron spectroscopy, XRD: X-ray diffraction, XRF: X-ray fluorescence, OPC: optical particle counter, CPC: condensation particle counter, FTIR: Fourier-transform infrared spectroscopy, NMR: nuclear magnetic resonance, ICP: inductively coupled plasma, AA: atomic absorption spectroscopy, HPLC: high-performance liquid chromatography, ESR: electron spin resonance, DLS: dynamic light scatter, EDS: X-ray energy dispersive spectroscopy, AES: Auger electron spectroscopy, TGA: thermogravimetric analysis, LAL: Limulus amoebocyte lysate, FCS: fluorescence correlation spectroscopy, EPR: electron paramagnetic resonance, TEM: transmission electron microscopy, SEM: scanning electron microscopy, BET: Brunauer-Emmett-Teller, pH: hydrogen-ion activity, OES: optical emission spectroscopy, SPM: single potentiometric method, PCM: potentiometric conductometric method, SPR: surface plasmon resonance, FP: first principles, MIP: mercury intrusion porosimetry, NBT: nitroblue tetrazolium test, MALD-TOF: matrix-assisted laser, desorption/ionization-time of flight, EELS: electron energy loss spectroscopy, and TET: TNF-α expression test (where: TNF is tumor necrosis factor).

Table 1.
Nanomaterial evaluation: main relevant properties related to risk assessment and usual material characterization techniques [16–20].

Concentration, type/chemical, and physical state, in general, are the most important technical aspects related to HSE management of workplace under evaluation.

A careful analysis on nanomaterial's processes such as handling, storage, manipulation, production, and incorporation are essential to determine exposure routes and to identify possible emission procedures or other safety aspects. **Figure 3** details the possible relations between biokinectics and pathways for nanoparticles exposure route and translocation .

The interaction between nanoparticles and cells is important to determine health surveillance (target organs and excretory pathways) and the possible damage mechanics and effects related to nanotoxicity such as physical damage of lysosomes, lipid peroxidation in vesicles, DNA damage of cellular nucleus, mitochondrial damage of cell mitochondria, protein misfolding on Golgi apparatus, and cellular membrane damage from oxidative stress, surfactant interaction, membrane damage by ions, and disruption of cell membranes [22]. Then, new properties from nano-sized effects (high area and energy, specific functionalization and quantum effects) provide different biological, physical, and/or ecological effects, turning risk assessment and evaluation more complex, sometimes more specific and time-consuming.

3.1 Occupational nanoparticle exposure routes and controls

Occupational exposure routes are intimately related to physical state of nanoparticles from processing methods (synthesis), processing equipment, workplace design, pollution-control equipment, handling steps, and operational procedures. Mainly, dispersion and inhalation are the main exposure contamination route. HEPA class filter are very efficient to control the number of nanoparticles disperse on air (>99.97%), while ULPA class filters have at least 99.995% of efficiency [22]. Use of a proper project design of air filtration is one of most important stages to avoid personal and environmental contamination. **Table 2** shows the typical HEPA and ULPA filter specifications based on ISO 1822:2009 (*) and the principal four primary mechanisms of filter particle collection [23]. The use of clean rooms or lamellar flow chamber or environmental/gas-controlled glove box could be applied to contain particulates during the handling of nanomaterials [24]. Personal protective equipment (PPE) also uses HEPA filtration on mask or coupled filtration devices. Main aspects of contamination and failure to prevent contamination regarding filtration masks came from poor personal training to fix and use

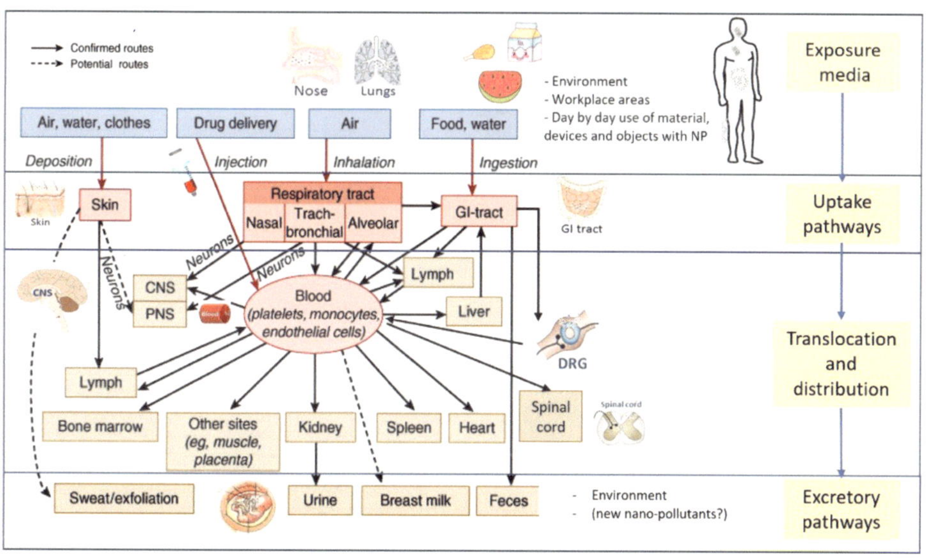

Figure 3.
Biokinectics of nanoparticles related to exposure routes, uptake pathways, translocation/distribution and excretory pathways of NP (adapted from Ref. [21]).

HEPA/ULPA Class	Retention (total %)	Filtration Mechanisms	Main Capture Sizes (µm)
H13	> 99.95	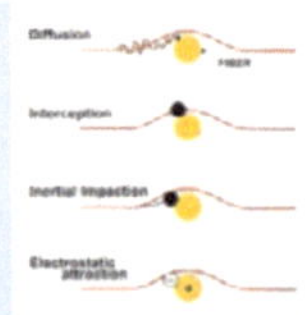	- Diffusion: < 0.1 microns - Interception: ~0.4 microns - Inertial impaction: ~0.4 microns - Electrostatic attraction: depend on size, charge and distance between particles and filter treated fibers.
H14	> 99.995		
U15	> 99.9995		
U16	> 99.99995		
U17	> 99.999995		
HEPA: High Efficiency Particulate Air ULPA: Ultra Low Particulate Air			
Retention efficiency based on MPPS (Most Penetrating Particle Size) test using 300 nm size particles.			

Table 2.
HEPA and ULPA filters' retention characteristics.

masks and the incorrect size and face matches (respiration facepiece and face). A chartered safety and health practitioner's evaluation based on work activities, operational procedures, workplace environment, nanoparticulate type/chemical concentration, and respirator fit factor should be applied according to occupational standards and guides [25–27].

The second most important exposure route of nanoparticles is from deposition or contact with skin. Usually, "perfect" skins without damages, cuts, or scratches work as barrier against nanoparticle penetration. However, the adequate gloves' and protective clothes' (made by high-density polyethylene fibers) specification should take in care several parameters such as glove material's chemical resistance compatibility [28], glove thickness, type/time of glove use (disposable/unsupported), temperature of use, type of labor activity and type, concentration, and agglomeration state of nanoparticles regarding the nanoparticle penetration behavior [29].

3.2 Nanotoxicity and information of the material safety data sheets (MSDS)

The material safety data sheets (MSDS) have essential information's and, regularly, they are used during the risk assessment and in workplaces as safe procedures

	OEDC (nano)	UN-CGH (chemicals)
Physicochemical properties	Melting point, boiling point, vapor pressure, K octanol/water, K organic/C/water*, water solubility	Appearance (physical state, color, etc.), upper/lower flammability or explosive limits, odor, vapor pressure, odor threshold vapor density, pH, relative density, melting point/freezing point, solubility(ies), initial boiling point and boiling range, flash point, evaporation rate, flammability (solid, gas), partition coefficient: n-octanol/water, auto-ignition temperature, decomposition temperature, and viscosity.
Environmental fate	Biodegradation, hydrolysis, atmospheric oxidation, bioaccumulation*	(Nonmandatory): • Results of tests of bioaccumulation potential, making reference to the octanol-water partition coefficient (K_{ow}), bioconcentration factor (BCF) • Other adverse effects (e.g., environmental fate, ozone layer depletion potential, photochemical ozone creation potential, endocrine disrupting potential, and/or global warming potential).
Ecological effects	Acute fish toxicity, acute Daphnia toxicity, alga toxicity, long-term aquatic toxicity, terrestrial effects	(Nonmandatory): • Acute or chronic aquatic toxicity data for fish, algae, crustaceans, and other plants, toxicity data on birds, bees, plants, etc.

	OEDC (nano)	UN-CGH (chemicals)
Human health effects	Acute oral toxicity, acute inhalation toxicity Akin irritation/corrosion Eye irritation/corrosion Repeated dose Genotoxicity (in vitro) Genotoxicity (in vitro, nonbacterial) Genotoxicity (in vivo) Reproductive toxicity Developmental toxicity Carcinogenicity* Organ toxicity (hepatotoxicity, cardiotoxicity, nephrotoxicity, etc.).	Information on the likely routes of exposure (inhalation, ingestion, skin, and eye contact). The SDS should indicate if the information is unknown: description of the delayed, immediate, or chronic effects from short- and long-term exposures. The numerical measures of toxicity (e.g., acute toxicity estimates such as the LD50 (median lethal dose)) – the estimated amount [of a substance] expected to kill 50% of test animals in a single dose. Description of the symptoms. This description includes the symptoms associated with exposure to the chemical including symptoms from the lowest to the most severe exposure. Indication of whether the chemical is listed in the National Toxicology Program (NTP) Report on Carcinogens (latest edition) or has been found to be a potential carcinogen in the International Agency for Research on Cancer (IARC) Monographs (latest editions) or found to be a potential carcinogen by OSHA.

From UN-GHS (classification and labeling of chemicals of UN-globally harmonized system).
Remark: the US National Cancer Institute published on this website several assays that have been standardized to work with a variety of nanomaterials. Information available at [38]. *Non-screening information data set (SIDS) endpoints.*

Table 3.
OECD endpoint test guidelines for nanomaterials and MSDS (material safety data sheet) parameter.

for hazardous chemicals and its mixtures. The SDS enable anticipation, evaluation, recognition, and control of workplace exposures and environmental hazards [30, 31]. The major problem of SDS of nanomaterials is the lack of information and/or wrong use of "bulk material" properties instead of nano. Evaluation of nano-SDS indicated that 35% of the sheets are unreliable [32]. About this issue, American Industrial Hygiene Association (AIHA) has a program that recognizes chemical hazard communication and environmental health professionals who specializes in authoring safety data sheets and labels [33]. Recently, SECO, a Swiss State Secretariat for Economic Affairs (SECO), published a guideline and two examples of synthetic nanomaterial's SDS [34, 35].

The knowledge of nanomaterial (quantitative) structure-activity relationships (Q)SAR is another important point of HSE evaluation during the risk assessment. Validated endpoints could explicate more clearly when a nanomaterial would be classified as "toxic" or "nontoxic" in particular exposure conditions providing more analytical, rational, and explicit data interpretation. **Table 3** shows the most common endpoints based on OECD test guidelines for nanomaterials and the UN-GHS. Notice that the UN-CGH for chemicals (including nanosized) just shows the parameter, while the OECD guideline could indicate and classify them [36, 37].

4. Risk assessment of carbon nanomaterials

Risk assessment could be defined when risk analysis and risk evaluation are carried in a joint process [39]. Health Safety Executive from the United Kingdom (HSE-UK) uses a simple five-step risk assessment based on the following topics:

- Identify the hazard.
- Decide who might be harmed and how.
- Evaluate the risks and decide on precautions.
- Record your findings and implement them.
- Review your assessment and update if necessary.

The application of this kind of assessment can be useful; however, it needs some adjustments due to uncertainties and lack of information (e.g., long-term or chronic evaluation of nanomaterials exposure), new materials without completed toxicological effects/studies or accurate dose-response database (OEL—occupational exposure limits, TLV—threshold limit values, OSHA/PEL—occupational safety and health administration-permissible exposure limits, WEEL—workplace environmental exposure level, NIOSH/PEL—National Institute for Occupational Safety and Health-recommended exposure limit, IOELV—indicative occupational exposure limit value, etc.) [40].

The availability of occupational and epidemiological data for chemicals and nanomaterials is a key aspect of risk assessment. The amount of new chemicals produced and released on the market is huge and huge means a hundred thousand per year. Based on Chemical Abstract Services (CAS) registry (a division of American Chemical Society), since 1800s, more than 145 million organic and inorganic substances were disclosed in the literature [41], while the publication 2018 TLV and BEIs from the American Conference of Governmental Industrial Hygienists (ACGIH) presents around 700 chemicals with TLV-STEL or TLV-TWA values (short-term exposure limit and time-weighted average, over the 8-hour working day). At the nanoworld, only 56 nanomanufactured materials have occupational exposure limit (OEL) proposed values [42]. **Table 4** presents some of the proposed

Types of nanocarbons	Exposure	Reference
Multiwalled carbon nanotubes (MWCNT)	No effect concentration in air <2.5 μg/m³	Luizi, 2009 (from Nanocyl)
Multiwalled carbon nanotubes (MWCNT)	Occupational exposure limit (OEL) < 50 μg/m³ for 8-hour TWA during a 40-hour workweek	Pauluhn, 2010 (from Bayer)
Fullerenes (C60)	Indicative no-effect level (INEL) < 7.4 μg/cm³	Aschberger, 2011
Multiwalled carbon nanotubes (MWCNT) 10 nm	Indicative no-effect level (INEL) < 1.0 μg/cm³	Aschberger, 2011
Carbon nanotubes (CNTs)	Proposed nanoreference values (NRV) < 0.01 fibers/cm³	van Broekhuizen, 2012
Carbon nanotubes (CNTs)	Recommended exposure limit (REL) < 1.0 μg/m³ for 8-hour TWA during a 40-hour workweek	NIOSH, 2013 [64]
Carbon nanofibers (CNFs)	Occupational exposure limit (OEL) <0.01 fibers/cm³	Stockmann-Juvala, 2014
Carbon nanotube group, SWCNT, DWCNT, MWCNT	Occupational exposure limit (OEL) < 30 μg/m³ for 8-hour TWA during a 40-hour workweek	Nakanishi, 2015

Table 4.
Proposed exposure limit values for carbon nanomaterials.

exposure limit values for carbon nanomaterials. The suggested values vary enormously, and no consensus exists between the authors.

In this scenario, the adoption of a precautionary posture should be the only option. In its strict form, the precautionary principle "requires inaction when action might pose a risk," or in an active form, "it requires to choose the less risky alternatives when they are available, and for taking responsibility for potential risks" [43–45]. This should be adopted and carefully evaluated on control of risks (identification, evaluation, elimination, mitigation, monitoring, and communication) and on the choice of mitigation procedures when dealing with nanomaterials. The adoption of a hierarch for health and safety controls should be mandatory. **Figure 4** shows health, safety, and environmental hierarchy controls in a schematic representation.

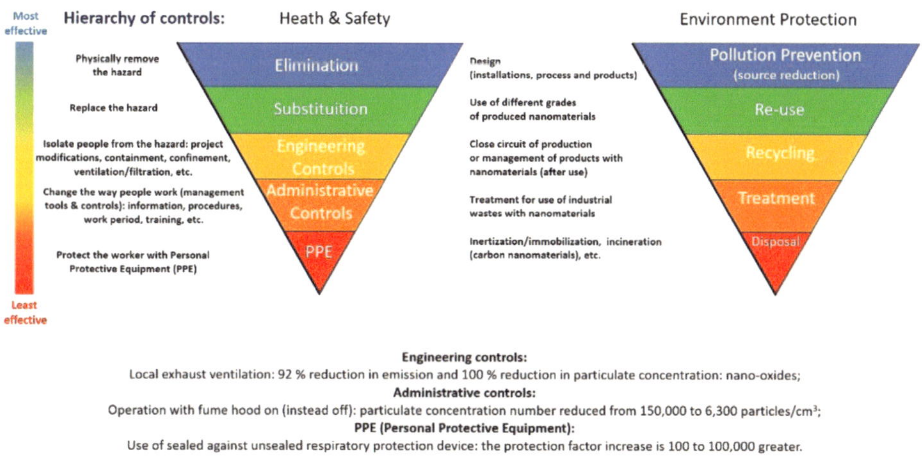

Figure 4.
HSE hierarchy controls (adapted from Ref. [46, 47])

Risk Management Involving NOAA's

Risk Identification: Synthesis, processing (dispersion, stabilization, functionalization and integration), handling, storage, product incorporation, use/end-use, disposal/treatment (inertization, immobilization, etc.)

Risk Analysis:
- Planning;
- Scope and boundary condition definitions;
- Identify: threats & hazardous events;
- Determine causes & frequency;
- Develop accidents scenarios
- Select relevant and typical accidents/contamination scenarios (prioritization);
- Determine/simulate/estimate consequences of accidents or contamination events;
- Determine/simulate/estimate) frequencies of accidents or contamination events;
- Assess the sensitivity of analysis;
- Establish and describe the risk pictures;
- Report the analysis.

Risk Evaluation:
- Evaluate the risk against risk acceptance criteria;
- Suggest and evaluate potential risk-reducing measures & training.

Monitoring & review
- **Hazard Identification:**
- **Hazard characterization**
- **Exposure assessment**

Risk Treatment: Prevention & Mitigation
- Eliminate, substitute and or minimize;
- Prevent;
- Detect and warn
- Mitigate.

Risk Communication & risk consultation

Figure 5.
Risk management of NOAA's topics of risk identification, analysis, evaluation, treatment, and communication.

Organization (acronym)/country and website
IRSST (Canada), Institut de recherche Robert-Sauve en sante et en securite du travail/www.irsst.qc.ca
ILO/ONU (International Labour Office-Swiss)/www.ilo.org
CDC-NIOSH and OSHA (USA), Centers for Disease Control and Prevention—The National Institute for Occupational Safety and Health/https://www.cdc.gov/niosh/topics/nanotech/pubs.html Occupational Safety and Health Administration/https://www.osha.gov/dsg/nanotechnology/nanotech_standards.html
Cordis (Europe), Community Research and Development Information Service/www.cordis.europa.eu/nanotechnology
SWA (Australia), Safe Work Australia/http://www.safeworkaustralia.gov.au/https://www.worksafe.qld.gov.au
ISO (Switzerland), International Organization for Standardization/www.iso.org
HSE (UK), Health and Safety Executive/www.hse.gov.uk, www.safenano.org
BAuA (Germany), German Federal Institute for Occupational Safety and Health/www.baua.de
DTU (Denmark), Danish Ecological Council and Danish Consumer Council/http://nanodb.dk/en/nanoriskcat/
ANSES (France), Agence Nationale de Securite Sanitaire de l'alimentation/www.anses.fr
AIST (Japan), National Institute of Advanced Industrial Science and Technology/https://www.aist.go.jp/index_en.html
OECD (France), The Organization for Economic Co-operation and Development/www.oecd.org/http://www.oecd.org/science/nanosafety/
RIVM (Holland), Dutch National Institute for Public Health and the Environment/https://www.rivm.nl/en/nanotechnology
ABDI (Brazil), Brazilian Agency for Industrial Development/www.abdi.com.br/http://livroaberto.ibict.br/handle/1/624
EUON (Europe), European Union Observatory for Nanomaterials/https://euon.echa.europa.eu/

Table 5.
Major protocols, standards, and guidelines on the safe handling of nanomaterials and the risks associated with nanoparticles available from different organizations [52].

Tool	Web address	Country
CB Nanotool	https://controlbanding.llnl.gov/	USA
IVAM Guidance	http://www.industox.nl/Guidance%20on%20safe%20handling%20nanomats&products.pdf	Holland
Swiss Precautionary Matrix	http://www12.esdc.gc.ca/sgpe-pmps/servlet/sgpp-pmps-pub?lang=eng&curjsp=p.5bd.2t.1.3ls@-eng.jsp&curactn=dwnld&pid=61345&did=5245	Swiss
Stoffenmanager Nano	https://nano.stoffenmanager.com/	Nederland
ANSES CB Tool	https://www.anses.fr/en/system/files/AP2008sa0407RaEN.pdf	France
NanoSafer CB	http://nanosafer.i-bar.dk/Default.aspx	Denmark
NCBRA—Nanomaterial Control Banding Risk Assessment	https://www.worksafe.qld.gov.au/data/assets/pdf_file/0005/110975/nanotechnology- control-banding-tool.pdf	Australia
CBG—Control Banding Guideline	http://www12.esdc.gc.ca/sgpe-pmps/servlet/sgpp-pmps-pub?lang=eng&curjsp=p.5bd.2t.1.3ls@-eng.jsp&curactn=dwnld&pid=61345&did=5245	Canada

Table 6.
Major risk assessment tools based on control banding and risk matrices ([53] and Google searching tool).

The most effective control is the elimination of the hazard, and the less effective is the use of protective equipment. The viability to apply each type should be analyzed, case-by-case, considering the identified risks and its classification.

The risk management of nanomaterials nano-objects, aggregates and agglomerates (NOAA) could be based on international standards like ISO 31000 [48] and should be incorporated in the quality, health, safety, and environmental system management. Adaptations for the correct use of risk assessment tools to deal with specificities and specific properties of nanomaterials are necessary [49, 50]. **Figure 5** shows the core points of risk management and several tools that could be used to facilitate the understanding and the day-by-day procedures.

The applying of safety aspects in early stage of projects involving nanomaterials is fundamental to mitigate risks and improve the quality (robustness) of HSE system. The CDC/NIOSH guide from Nanotechnology Research Center (NTRC): “Controlling Health Hazards When Working with Nanomaterials: Questions to Ask Before You Start” is a very good starting point to deal with nanomaterials in a safe manner [51]. **Table 5** shows major protocols, standards, and guidance on the safe manipulation of NOAAs.

4.1 Risk analysis and evaluation

Due to lack of exposure limits and lack of information concerning toxicological effects of nanomaterials (in general and carbon nanomaterials specifically), the use of risk assessment based on control banding has been widely applied. Several tools based on this concept are available as shown in **Table 6**.

5. Environmental risk management

Along with all carbon nanomaterial developmental stages—from laboratory to industrial scale operations—different routes can lead to environmental emissions. Decision making related to waste disposal and control actions to avoid (or allow) air and water emissions should be taken in the early stages, prior to the enhancement of production scales. These types of decision will also be needed after the use phase of the nano-enabled products. Naturally, through several possible paths, different environment compartments will interact (or better, are interacting) with carbon nanoparticles.

5.1 Air

Depending on the scale, type of synthesis route, and activity, carbon nanoparticle emissions to the atmosphere can occur in larger or smaller amounts. As mentioned, exposition via inhalation is the main concern in the field since several authors have reported possible effects and damage to the respiratory system [18, 54]. In this sense, important precautionary actions comprise the establishment of emission control strategies and exposure monitoring.

5.1.1 Emission control strategies

Nanoparticles in the air were traditionally referred to ultrafine particles [55]. Main processes can occur with these particles in the air: photochemical reactions, agglomeration, and deposition. The deposition depends on the gravitational settling velocity, which is proportional to the diameter of the particle—smaller nanoparticles in the air will deposit at a much slower rate than larger particles [56, 57].

It is possible to assume that the particle diffusion in the atmosphere is governed by Brownian motion, being the diffusion rate inversely proportional to the particle diameter. In that sense, it is possible to assume that "aerosolized" nanoparticles in an agglomerate form, even at a low mass concentration, will be deposited at a faster rate than larger particles [56–58].

Also, the specific properties of the nanomaterial such as surface charge, aspect ratio, and presence of functional groups or other molecules in the nanoparticle surface will play a significant role in the atmospheric processes, interfering in the photochemical reactions, agglomeration, and deposition. The behavior of these particles in the air is a complex and dynamic process.

Being detected the risk of a carbon nanomaterial becoming airborne, some control measures should be taken. Handling nanomaterials should be performed in a closed system, for instance, a glove box. If activities outside a closed system cannot be avoided, as during refilling or filling, dust should be extracted at the source, using exhaustion devices (like flexible air ducts). It is recommended by several agencies to use HEPA filters of class "H14" in safety cabinets or other exhaustion fume hoods at the workplace. The use of safety cabinets and microbiological safety cabinets, which recirculate air from the cabinet interiors, through a HEPA filter, back into the laboratory, can be used for small quantities of carbon nanomaterials in the absence of hazardous vapors or gases. The filter must be HEPA, and charcoal filters alone must not be used in this case. Also, suitable protective measures to avoid energetic processes that might generate airborne dusts or aerosols and the immediate sealing of bottles/vessels after use are simple and effective actions [22–26].

5.1.2 Exposure monitoring

Different methods can be used for measuring airborne nanomaterials. As illustrated in **Figure 6**, atmospheric particulate matter can be analyzed through different methods using passive or active sampling methodologies.

Online measurement methods are equipment-based measurements with low response time, giving in real time, the concentration of particles and nanoparticles in the air. Normally, it is a general measurement that cannot distinguish the type of nanomaterial (related to its composition or morphology) being analyzed. As for the carbon nanomaterials, the morphology is a differential property; the simple measurement of concentration cannot determine the presence of the nanomaterial in the air. Recently, more relevant indicators have emerged for describing nanoparticle aerosols, including particle number, surface and mass concentrations, and criteria relating to their size or optical properties (Raman-based equipment) [59, 60].

Off-line methods allow quantitative determination (by mass), and the identification of the type of nanomaterial, however, requires long-term sampling and the sampling equipment is often expensive. Since these methods comprise sample collection, the advantage is the possibility to analyze the air-collecting filters in order to determine the composition and to verify the presence of carbon nanoparticles in the filter. The complexity of these samples in terms of composition and the likely presence of other carbonaceous materials (amorphous carbon from diesel combustion, for instance) is a special challenge related to the air monitoring of carbon nanoparticles. It is important to say that quantification is another difficult task. Carbon nanoparticles are very light, and the gravimetric method gives a result related to the total particulate sampled. A recent study analyzed the use of thermogravimetric analyses in order to detect and differentiate different types of nanocarbon directly in the air-collecting filter [61]. For the off-line gravimetric method, conventional sampling regulations such as NIOSH Method 5040, NIOSH 0500, and NIOSH 0600 can be used [62, 63].

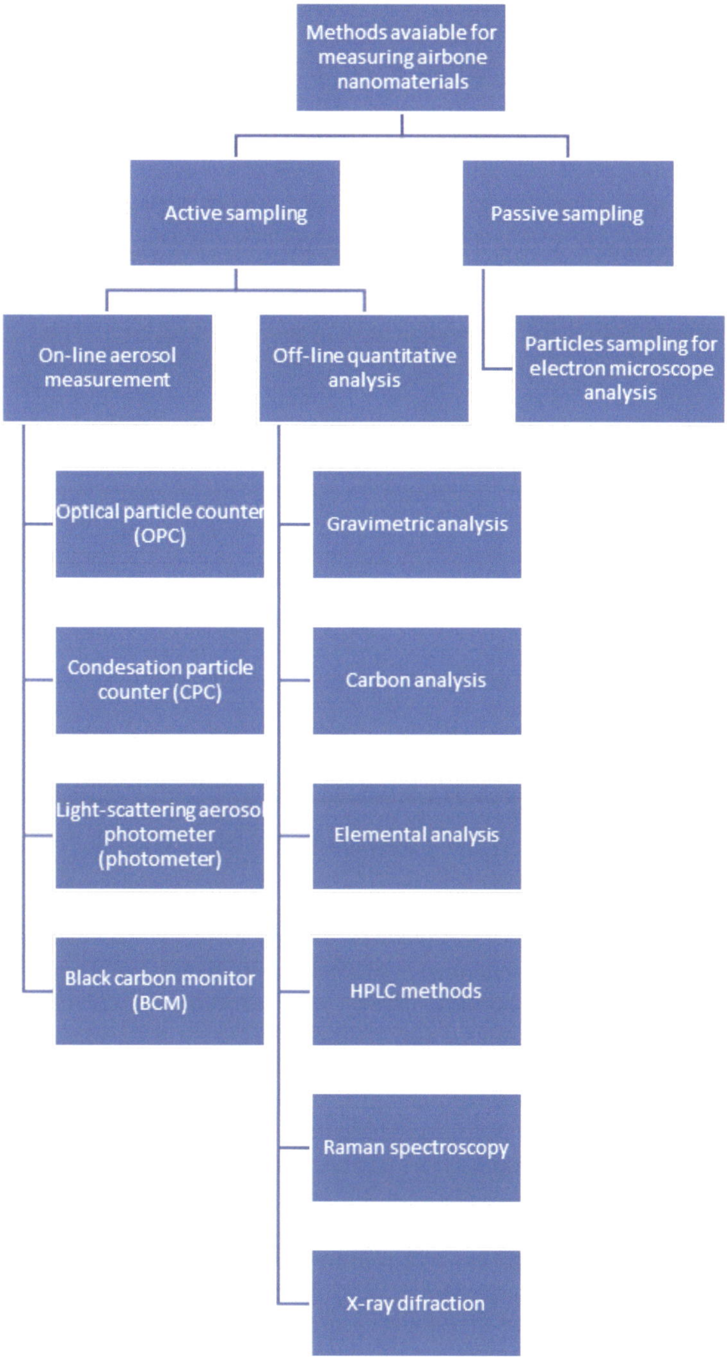

Figure 6.
Flowchart of methods for measuring airborne nanomaterials.

Passive methods are related to the natural deposition of those particles in different substrates (TEM grids, silica, holey carbon tapes) aiming to visualize its presence using an electron microscope. Although it allows identification of the carbon nanomaterial and its morphology, it is an expensive and time-consuming method. Also, the uncertainties related to the behavior of these particles in the air make it difficult to assume the time needed for its deposition and settling in those substrates.

5.2 Water and soil

Graphene and carbon nanotube water dispersions are a market reality. Although the fact of being, originally, insoluble materials, the possibility to have water dispersions containing carbon nanomaterials is often considered, since the delivery of a product in a liquid media can allow its use for different application purposes.

Different types of liquids containing carbon nanomaterials can be generated in research and (pre) industrial facilities. The use of acids and surfactants in synthesis and functionalization routes are common and, therefore, will be components of the produced liquid waste. Besides the nanomaterial, the presence of other substances must be evaluated since the knowledge of their own risk and best disposal practice must be considered.

Carbon nanoparticles may interact with aquatic systems by sewer discharges, or indirectly, coming from soils during the solid nanomaterial (or products containing the nanomaterial) disposal in landfills.

It is unlikely that public water treatment systems will be able to remove those particles from the influent. Specific properties such as surface charge and the presence of surfactants are some of the difficulties related to the removal of these nanomaterials in a general way [65, 66]. In some cases, at higher efficiency regime, the nanomaterials are removed from the influent and adsorbed into the sludge that will be disposed in the soil [67].

While there are no specific rules that regulate the disposal of nanomaterials in the environment, it is necessary to create a precautionary strategy for the best environmental management.

A possible approach is the treatment of the liquid nanowaste, aiming the removal of the nanoparticles from the water and its possible reuse in the production process. Liquid waste treatment processes are established in the industry, making possible a similar development in the nanotechnology field. The tendency of some nanocarbons, like graphene nanosheets, to aggregate and form a precipitated agglomerated due to π–π staking interactions, can facilitate this separation [68]. On the other hand, graphene nanosheets with attached metal nanoparticles easily aggregate in the presence of ions [66]. The treatment and recycling are strict approaches that eliminate the risk of any possible harmful effects to aquatic systems, since the nanoparticle will not reach the environment. Although it seems an interesting choice, the establishment of treatment/recycling routes will require research and development efforts, involving nontrivial issues related to characterization of nanocarbon on complex matrices and a clear understanding of key nanomaterial properties.

Besides the challenge of such approach, it has clear advantages since it adds to the nanomaterial development of a safe-by-design practice in accordance with sustainable principles, therefore creating the possibility of specific solutions for each case. A process to recover single-walled carbon nanotube anode from battery anodes [69] and the use of serpentine to treat waste graphene were already reported [70] as examples of recycling the nanocarbons used in specific applications. It is a complex subject with an urgent need for recycling and end-of-life (EOL) studies. Specifically, the generation of the nonproduct outputs in nanocarbon production process is common [71]. The industrial scale production highlights this need: the projected global annual production of carbon-based nanocarbons was in order of a few hundred tons in 2001 and estimates 58,000 tons annually to 2020 [72].

According to several recommendations [73–76], the incineration is another method to treat these types of waste. These recommendations are based on the precautionary principle, and independent of the amount, the presence of nanomaterials in liquids that should be discarded lead to its classification as hazardous wastes.The Organization for Economic Co-operation and Development (OECD) [74], in 2013, published a document providing an overview of scientific findings on the behavior and exposure of engineered nanomaterials (ENMs) during the waste incineration process in order to identify knowledge-gaps regarding specific aspects of the disposal of waste containing nanomaterials (WCNMs). According with this document, all waste incineration plants should be equipped with a flue gas treatment system as, for instance, described in the European Union BREF document (Integrated Pollution Prevention and Control—Reference Document on the best available techniques for Waste Incineration, August 2006). Until today, there are still a large number of waste incineration plants worldwide that do not have adequate flue gas treatment systems. In addition, the treatment and disposal of solid residues from waste incineration also require further research that should include the determination of conditions that enable the efficient removal of nanocarbons from the solid waste. Each material needs to be evaluated in an individual way. Decomposition temperatures depend on the nanomaterial-specific properties: size, shape, degree of defects, and presence or not of functional groups can interfere on it.

Besides the liquids containing nanocarbons, solid wastes are also generated in research facilities and in industrial settings producing or working with carbon nanomaterials. In terms of risk management, an important question regarding the form of the nanomaterial in the solid waste: Is it bound in a matrix or free in a powder form? The classification of the types of solid and liquid wastes is a fundamental step. A careful look in the processes and the development (and clear communication) of disposal practices are essential to avoid unnecessary emissions to the atmosphere or the disposal of a solid nanomaterial, in powder form in landfill soil.

Landfill waste disposal of carbon nanomaterials leavings derived from production facilities and from nano-enabled products are relevant routes leading to the soil environment. Another possibility is the use of nanoproducts directly in the soil for agricultural and construction purposes—cement containing carbon nanotubes is a well-studied application [77].

Progress in nanowaste management also requires studies on the environmental impact of these new materials. Knowing that several parameters must be evaluated when assessing the risk—such as the amount of nanomaterial to be discarded and its singular properties—there is a preponderant factor: the complexity of the natural environment, which makes it difficult to carry out systematic studies for the identification of the risk. Studies focusing the identification of the harmful effects of these nanomaterials in realistic scenarios are rare. Works modeling the flows of engineered nanomaterials during waste handling are the important ways to support this lack of information [78, 79].

6. Conclusions and final comments

Nowadays, the amount of information of HSE of nanomaterials including the new nanocarbons is still not completed; however, the correct use of risk assessment and risk analysis tools based on a precautionary vision could mitigate risks and control the exposure at workplaces. Improving the quality and ratability information of SDS and improving the grouping of nanomaterials by risk classes based on properties and biological effects must be a continuous work. A multidisciplinary environmental assessment and use of LCA tools to nanomaterials must be continued, improving risk models and taking care of epidemiological aspects of human and environmental interaction with manmade nanoparticles.

Acknowledgements

Authors would like to thank the MGgrafeno project, a joint initiative of Codemig, Center for Development of Nuclear Technology (CDTN) and Federal University of Minas Gerais (UFMG).

Author details

Guilherme Lenz e Silva[1,2*], Camila Viana[2], Danieli Domingues[2] and Fernanda Vieira[2]

1 Polytechnic School of Engineering, University of São Paulo, São Paulo, Brazil

2 MGgrafeno-CDTN/CNEN-Codemig, Belo Horizonte, Brazil

*Address all correspondence to: guilhermelenz@usp.br

References

[1] The Royal Society: RS/RAE. Nanoscience and Nanotechnologies: Opportunities and Uncertainties. London, UK: The Royal Society and the Royal Academy of Engineering; 2004. 113 p

[2] ETC Group. No Small Matter II: The Case for a Global Moratorium. Size Matters! Winnipeg, Canada: ETC Group; 2002. 20 p

[3] OECD/Allianz. Small sizes that matter: Opportunities and risks of nanotechnologies [Internet]. 2003. Available from: https://www.oecd.org/science/nanosafety/44108334.pdf [Accessed: 2019-01-10]

[4] European Commission. Nanotechnologies: A preliminary risk analysis on the basis of a workshop organized in Brussels on 1-2 March 2004 by the health and consumer protection directorate general of the European Commission [Internet]. 2004. Available from: https://ec.europa.eu/health/ph_risk/documents/ev_20040301_en.pdf [Accessed: 2019-01-10]

[5] Swiss Reinsurance Company [Internet]. Available from: https://www.nanowerk.com/nanotechnology/reports/reportpdf/report93.pdf [Accessed: 2019-01-10]

[6] Obersdörster GFJ, Oberdörster G, Penney DP, Soderholm SC, Gelein R, Piper HC. Increased pulmonary toxicity of ultrafine particles. 1. Particle clearance, translocation, morphology. Journal of Aerosol Science. 1990;**21**:381-384. DOI: 10.1016/0021-8502(90)90064-5

[7] Oberdörster G, Ferin J, Finkelstein G, Wade P, Corson N. Increased pulmonary toxicity of ultrafine particles. 2. Lung lavage studies. Journal of Aerosol Science. 1990;**21**:384-387. DOI: 10.1016/0021-8502(90)90065-6

[8] Hornyak T. Spotlight: Carbon belts and chains aid nanotech resurgence: Private-sector cash and technological innovation are helping Japan regain its lead in nanocarbon research. Nature. 2017;**552**:S45-S47. DOI: 10.1038/d41586-017-07451-x

[9] Kroto HW, Heath JR, O'Brien SC, Curl RF, Smalley RE. C60: buckminsterfullerene. Nature. 1985;**318**:162-163. DOI: 10.1038/318162a0

[10] Iijima S. Helical microtubules of graphitic carbon. Nature. 1991;**354**(6348):56-58

[11] Geim AK, Novoselov KS. The rise of graphene. Available from: https://arxiv.org/ftp/cond-mat/papers/0702/0702595.pdf [Accessed: 2019-02-18]

[12] Lenz e Silva Guilherme FB. Carbono e nanotecnologia. Metalurgia & Materiais. 2011;**67**:309-315

[13] European Commission Health & Consumer Protection Directorate-General. The appropriateness of existing methodologies to assess the potential risks associated with engineered and adventitious products of nanotechnologies. Available from: http://ec.europa.eu/health/ph_risk/committees/04_scenihr/docs/scenihr_o_003b.pdf [Accessed: 2019-02-18]

[14] NNI/United States National Nanotechnology Initiative. What is nanotechnology. Available from: https://www.nano.gov/nanotech-101/what/definition [Accessed: 2019-02-18]

[15] Lenz e Silva GF, Hurt R. Risk assessment of nanocarbon materials using a multi-criteria method based on analytical hierarchy process. Current Bionanotechnology. 2016;**02**:1-6

[16] Sutariya VB, Pathak Y. Biointeractions of Nanomaterials.

1st ed. CRC Press; 2014. ISBN: 9781466582385, Reference—480 p

[17] Tiede. Considerations for environmental fate and ecotoxicity testing to support environmental risk assessments for engineered nanoparticles. Journal of Chromatography A. 2009;**1216**(3):503-509. DOI: 10.1016/j.chroma.2008.09.008

[18] Bengt F et al. Safety assessment of graphene-based materials: Focus on human health and the environment. ACS Nano. 2018;**12**(11):10582-10620. DOI: 10.1021/acsnano.8b04758

[19] Anshika A, Raheja A, Saranath P, Upender G. Nanomaterial toxicity. Available from: http://www.dstuns.iitm.ac.in/teaching-and-presentations/teaching/graduate-courses/electron-spectroscopy/Presentations-2009/Write-up%202%20-%20Nanomaterial%20%20toxicity.pdf [Accessed: 2019-02-18]

[20] Buzea C, Blandino IP, Robbie K. Nanomaterials and nanoparticles: Sources and toxicity. Biointerphases. 2007;**2**(4):MR17-MR172

[21] Oberdorster G, Kane AB, Klaper RD, Hurt RH. Nanotoxicology. In: Klaassen CD, editor. Casarett and Doull's Toxicology—The Basic Science of Poisons. New York, NY, USA: McGraw Hill; 2013. pp. 1189-1229

[22] Perry Marshall JL, Agui JH, Vijayakumar R. NASA—Submicron and nanoparticulate matter removal by HEPA-rated media filters and packed beds of granular materials. Available from: https://ntrs.nasa.gov/archive/nasa/casi.ntrs.nasa.gov/20170005166.pdf [Accessed: 2019-02-18]

[23] Liu G, Xiao M, Zhang X, Gal C, Chen X, Liu L, et al. A review of air filtration technologies for sustainable and healthy building ventilation. Sustainable Cities and Society. 2017;**32**:375-396. DOI: 10.1016/j.scs.2017.04.011. ISSN: 22106707. Available from: http://centaur.reading.ac.uk/71549/1/A%20review%20of%20air%20filtrati on%20technologies%20for%20sustainable%20and%20healthy%20building%20v entilation.pdf [Accessed: 2019-02-18]

[24] CDC. Protecting workers during nanomaterial reactor operations. Available from: https://www.cdc.gov/niosh/docs/2018-120/pdf/2018-120.pdf [Accessed: 2019-02-18]

[25] 3M, #171: Nanotechnology and respirator use. Revised June, 2015. Available from: https://multimedia.3m.com/mws/media/376179O/nanotechnology-and- respirator-use.pdf [Accessed: 2019-02-18]

[26] Shaffer RE, Rengasamy S. Respiratory protection against airborne nanoparticles: A review. Journal of Nanoparticle Research. October 2009. DOI: 10.1007/s11051-009-9649-3

[27] OSHA. Hazard communication, hazard classification guidance for manufacturers, importers, and employers. Available from: https://www.osha.gov/Publications/OSHA3844.pdf [Accessed: 2019-02-18]

[28] Assurance hand protection—chemical resistant chart. Available from: http://webfiles.ehs.ufl.edu/Assurance.pdf [Accessed: 2019-02-18]

[29] Dolez P et al. Development of a method of measuring nanoparticle penetration through protective glove materials under conditions simulating workplace use. Available from: https://www.irsst.qc.ca/media/documents/PubIRSST/R-785.pdf?v=2019-01-23 [Accessed: 2019-02-18]

[30] Available from: https://www.aiha.org/government-affairs/PositionStatements/position05_MSDS-09-02-2005.pdf [Accessed: 2019-02-18]

[31] Available from: https://www.osha.gov/Publications/OSHA3844.pdf [Accessed: 2019-02-18]

[32] Hodson L, Eastlake A, Herbers R. An evaluation of engineered nanomaterial safety data sheets for safety and health information post implementation of the revised hazard communication standard. Journal of Chemical Health and Safety. 2018. DOI: 10.1016/j.jchas.2018.10.002

[33] AIHA. Available from: https://www.aiha.org/registries/lar/Pages/sds-label-authoring-registry.aspx [Accessed: 2019-02-18]

[34] Federal Department for Economic Affairs (FDEA), State Secretariat for Economic Affairs (SECO). Working conditions chemicals and occupational health—safety data sheet (SDS): Guidelines for synthetic nanomaterials. 2016. Available from: https://www.bag.admin.ch/dam/bag/en/dokumente/chem/nanotechnologie/sicherheitsdatenblatt-leitfaden-nano.pdf.download.pdf/sicherheitsdatenblatt-leitfaden-nano.pdf [Accessed: 2019-02-18]

[35] Federal Department for Economic Affairs (FDEA), State Secretariat for Economic Affairs (SECO). Two examples for the guidelines Safety data sheet (SDS) for synthetic nanomaterials. 2016. Available from: https://www.bag.admin.ch/dam/bag/en/dokumente/chem/nanotechnologie/sicherheitsdatenblatt-beispiele-leitfaden.pdf.download.pdf/sicherheitsdatenblatt-beispiele-leitfaden.pdf [Accessed: 2019-02-18]

[36] Deveau M et al. The global landscape of occupational exposure limits—Implementation of harmonization principles to guide limit selection. Journal of Occupational and Environmental Hygiene. 2015;**12**(suppl 1):S127-S144. DOI: 10.1080/15459624.2015.1060327. Available from: https://www.ncbi.nlm.nih.gov/pmc/articles/PMC4654639/ [Accessed: 2019-02-18]

[37] OECD. Guidance document on the validation of (quantitative) structure-activity relationships [(Q)SAR] models. Available from: www.oecd.org/officialdocuments/publicdisplaydocumentpdf/?doclanguage=en&cote=env/jm/mono(2007)2 [Accessed: 2019-02-18]

[38] https://ncl.cancer.gov/resources/assay-cascade-protocols

[39] Rausand M. Risk Assessment: Theory, Methods, and Applications. 1st ed. ISBN-13: 978-0470637647

[40] Mihalache R, Verbeek J, Graczyk H, Murashov V, van Broekhuizen P. Occupational exposure limits for manufactured nanomaterials, a systematic review. Nanotoxicology. 2017;**11**(1):7-19. DOI: 10.1080/17435390.2016.1262920

[41] CAS content. The world's largest collection of chemistry insights. Available from: https://www.cas.org/about/cas-content. [Accessed: 2019-02-18]

[42] American Conference of Governmental Industrial Hygienists (ACGIH). 2018 TLVs and BEIs. Available from: https://www.acgih.org/ [Accessed: 2019-02-18]

[43] Available from: http://crnano.org/precautionary.htm [Accessed: 2019-02-18]

[44] Available from: http://www.scielo.br/scielo.php?pid=S1983-80422015000200244&script=sci_arttext&tlng=en [Accessed: 2019-02-18]

[45] Elliott KC. Nanomaterials and the precautionary principle. Environmental Health Perspectives. 2011;**119**(6):A240. DOI: 10.1289/ehp.1103687

[46] https://www.cdc.gov/niosh/topics/hierarchy/default.html

[47] Hierarchy controls. http://www.sun-fp7.eu/wp-content/

uploads/2017/01/SUN-risk-management-guidelines.pdf

[48] ISO 31000:2018-ISO 31000:2018. Risk management—Guidelines

[49] https://www.rivm.nl/bibliotheek/rapporten/2014-0157.pdf

[50] Guidance on available methods for risk assessment of nanomaterials. http://www.invassat.gva.es/documents/161660384/162311778/01+Guidance+on +available+methods+for+risk+assesment+of+nanomaterials/8cae41ad-d38a-42f7-90f3-9549a9c13fa0

[51] https://www.cdc.gov/niosh/docs/2018-103/pdfs/Nano_MP2_2018-103_508.pdf?id=10.26616/NIOSHPUB2018103

[52] Lenz e Silva GFB, Lenz e Silva LC. Occupational health and safety: Reflection on potential risks and the safety handling of nanomaterials. In: Vigilância Sanitária Em Debate: Sociedade, Ciência & Tecnologia. Vol. 1. 2013. pp. 42-51

[53] Liguoria B, Hansen SF, Baun A, Jensen KA. Control banding tools for occupational exposure assessment of nanomaterials—Ready for use in a regulatory context? NanoImpact. 2016;**2**:1-17. DOI: 10.1016/j.impact.2016.04.002

[54] Poulsen SS et al. MWCNTs of different physicochemical properties cause similar inflammatory responses, but differences in transcriptional and histological markers of fibrosis in mouse lungs. Toxicology and Applied Pharmacology. 2015;**284**:16-32

[55] Lead JR, Wilkinson KJ. Environmental colloids: Current knowledge and future developments. In: Wilkinson KL, Lead JR, editors. Environmental Colloids: Behaviour, Structure and Characterization. Chichester, UK: John Wiley; 2006. pp. 1-15

[56] Hartmann NIB, Skjolding LM, Hansen SF, Baun A, Kjølholt J, Gottschalk F. Environmental Fate and Behaviour of Nanomaterials: New Knowledge on Important Transfomation Processes. Copenhagen K: Danish Environmental Protection Agency; 2014 Environmental Project No. 1594

[57] Friedlander SK, Pui DYH. Emerging issues in nanoparticle aerosol science and technology (NAST). NSF Workshop Report. June 27-28; Los Angeles: University of California; 2003. p. 15

[58] Aitken R, Creeley C, Tran C. Nanoparticles: An occupational hygiene review. Institute of occupational medicine for the health and safety executive. Edinburgh, UK: 2004

[59] Ogura I. Guide to Evaluating Emission and Exposure of Carbon Nanomaterials (Carbon Nanotubes and Graphenes). 2018. Available from: https://www.researchgate.net/publication/324855977_Guide_to_Evaluating_Emission_and_Exposure_of_Carbon_Nanomaterials_carbon_nanotubes_and_graphenes

[60] Stat Peel. [Internet]. Available from: https://www.statpeel.com/. [Accessed: 2019-02-16]

[61] Viana CO, Domingues DS, Nascimento JP, Vieira F, Fernandes TFD, Silva DEL, et al. Graphene detection in air: A proposal based on thermogravimetric behaviour. Journal of Physics Conference Series. 2019 forthcoming

[62] NIOSH. NIOSH Method 5040: Issue 3. Diesel particulate matter (as elemental carbon). 2003. Available from: http://www.cdc.gov/niosh/docs/2003-154/pdfs/5040.pdf

[63] NIOSH. Nanotechnology, 10 Critical Topic Areas. NIOSH Publication 2009-125. 2011. 104 p. Available from: http://www.cdc.gov/niosh/docs/2009-125/pdfs/2009-125.pdf/http://www.cdc.gov/niosh/topics/nanotech/critical.html

[64] NIOSH. Occupational exposure to carbon nanotubes and nanofibres. Current Intelligence Bulletin. 2013;**65**. Available from: https://www.cdc.gov/niosh/docs/2013-145/pdfs/2013-145.pdf

[65] Musee N. Nanowastes and the environment: Potential new waste management paradigm. Environment International. 2011;**37**:112-128. DOI: 10.1016/j.envint.2010.08.005

[66] Zhang Y, Chen Y, Westerhoff P, Hristovski K, John C, Crittenden JC. Stability of commercial metal oxide nanoparticles in water. Water Research. 2008;**42**:2204-12.7

[67] Limbach LK, Bereiter R, Muller E, Krebs R, Galli R, Stark WJ. Removal of oxide nanoparticles in a model wastewater treatment plant; influence of agglomeration and surfactants on clearing efficiency. Environmental Science & Technology. 2008;**42**(15):5828-5833

[68] Nam B, Lee H, Goh H, Lee YB, Choi WS. Sandwich-like graphene nanocomposites armed with nanoneedles. Journal of Materials Chemistry. 2012;**22**:3148-3153

[69] Schauerman CM, Ganter MJ, Gabrielle G, Babbitt Callie W, Raffaelle Ryne P, Landi Brian J. Journal of Materials Chemistry. 2012;**22**:12008. DOI: 10.1039/c2jm31971c

[70] https://patents.google.com/patent/CN104192850A/en https://www.sciencedirect.com/science/article/pii/S0008622313006891

[71] Reijnders L. Safe Recycling of Materials containing Persistent Inorganic and Carbon Nanoparticles. Netherlands: University of Amsterdam. 2014; DOI: 10.1533/9780857096678.3.222

[72] Royal Society and Royal Academy of Engineering Report on Nanotechnology. Nanoscience and Nanotechnologies: Opportunities and Uncertainties. The Royal Society and Royal Academy of Engineering; 2004

[73] ENV/EPOC/WPRPW(2013)3/FINAL 30 Oct. 2015. Working Party on Resource Productivity and Waste—Incineration of Waste Containing Nanomaterials

[74] OECD. Nanomaterials in Waste Streams—Current Knowledge on Risks and Impacts. Paris: OECD Publishing; 2016. DOI: 10.1787/9789264249752-en

[75] Integrated Pollution Prevention and Control. Reference Document on the Best Available Techniques for Waste Incineration. Executive Summary. August 2006

[76] Mueller NC, Jelena B, Wang J, Andrea U, Nowack B. Modeling the flows of engineered nanomaterials during waste handling. Environmental Science: Processes & Impacts. 2013;**15**(1):251-259. DOI: 10.1039/C2EM30761H

[77] De P, Júnia N, Calixto JM, Ladeira LO, Ludvig P, Souza TCC. Tensile strength of oil-well cement pastes produced with carbon nanotubes directly synthesized on clinker. The International Journal of Engineering and Science. 2018;7:57-62

[78] Viana CO, Vaz RP, Cano A, Santos AP, Cançado LG, Ladeira LO, et al. Physiological changes of the lichen *Parmotrema tinctorum* as result of carbon nanotubes exposition. Ecotoxicology and Environmental Safety. 2015;**120**:110-116. DOI: 10.1016/j.ecoenv.2015.05.034

[79] Auffan M, Tella M, Santaella C, Brousset L, Paillès C, Barakat M, et al. An adaptable mesocosm platform for performing integrated assessments of nanomaterial risk in complex environmental systems. Scientific Reports. 2015;**4**:5608

Section 3

Environmental and Health Impact of Nanomaterials

Chapter 4

Oxidative Stress Produced by Urban Atmospheric Nanoparticles

Daniela-Rodica Mitrea, Alina-Mihaela Toader and Oana-Alina Hoteiuc

Abstract

In urban areas, the diesel-fuelled and bio-fuelled vehicles represent the major sources of nanoparticles complemented by nanotechnology with different types of particles, in addition to natural and to other anthropogenic sources. The atmospheric nanoparticles differ in composition, size, shape or oxidant capacity, presenting a large variability that causes difficulties in their measurements and health impact identification. The oxidative stress can be initiated by atmospheric nanoparticles through different mechanisms: interaction between nanoparticles and tissue cells, cellular internalisation of nanoparticles, activation of signalling pathways, decrease of the cellular antioxidants, activation of the pro-inflammatory cascade, lipid peroxidation, activation of cellular signalling pathway that leads to apoptosis, etc. Ultrafine particles (<100 nm) represent ~80% of the total atmospheric particles and produce inflammation through oxidative stress mechanisms. The atmospheric nanoparticles can penetrate the skin and can be inhaled or ingested affecting different organs and leading to different diseases: neurodegeneration, thrombogenesis, atherosclerosis, asthma, lung cancer, heart arrest, etc.

Keywords: nanoparticles, particulate matter, urban atmosphere, reactive oxygen species, oxidative stress, atmospheric ultrafine particles, diesel exhaust particles, air pollution

1. Introduction

Urban atmosphere contains a mixture of nanoparticles with high variation in structure, number and chemical composition. This atmospheric nanoparticles' diversity is in close relationship with the natural or industrial sources, urban area location and meteorological conditions. In the troposphere, the nanoparticles can be found at all altitudes from sea level to 10 km, with concentrations that can largely vary between 10^2 and 10^5 particles/cm^3 [1]. Atmospheric nanoparticles are of different chemical structures (unstable molecules or ions, stable nuclei) with various shapes (most of them have irregular aspect) and with many origins. The human body can be affected by nanoparticles exposure, because these particles penetrate easily through skin, respiratory system, or digestive tract. The atmospheric nanoparticles can produce oxidative stress through different mechanisms, from the reactive oxygen species (ROS) synthesis till the decreasing of body's antioxidant capacity [2].

This chapter will use the terms nanoparticles (NPs) or particulate matter ($PM_{0.1}$) for particles with sizes lower than 100 nm (lower than 0.1 μm).

2. Atmospheric nanoparticles

Earth's atmosphere contains particles with sizes between few nanometres and hundreds micrometres. They are produced naturally or by anthropogenic emissions, in concentrations that vary greatly, in accordance with the geographical areas and meteorological conditions. Naturally, the atmospheric particles can be produced from many sources: sea salt aerosols produced by breaking waves [3], cosmic dust [4], atmospheric formation of NPs from atmospheric particles (for example, photochemically induced nucleation) [5], forest's aerosols [6], volcanic eruptions, or forest fires. Anthropogenic emissions that produce atmospheric particles are represented by: vehicles, industry, power plants, incinerators, nanotechnology [7], fossil fuel exploitation, mining techniques, stone quarries [8], etc. Over the continents, close to the ground, the air contains 103–105 nanoparticles/cm^3, most of them chemically generated through coagulation or condensation [9].

Atmospheric particles are classified according to their diameter into: supercoarse, coarse, fine and ultrafine particles. Supercoarse particles (>10,000 nm diameter) do not enter the respiratory system. The coarse particles (diameter 2500–10,000 nm), represented by pollen, spores, sea salt aerosols, or by the particles generated through wind erosion, deposit on the ground within few hours after their production but can also enter the upper airways of the human respiratory tract, from where they can be eliminated back into the atmosphere by coughing, or they can be ingested by swallowing [10]. The fine particles (100–2500 nm diameter) produced naturally (through nuclei-mode particles coagulation or through vapour molecules condensation on particles surfaces) but mostly through anthropogenic processes (vehicles or industry emissions) [11], can enter the respiratory tract till the alveoli level. Because the fine particles have small diameters, the gravitational forces affect them only partially and they can persist in the atmosphere for days or weeks, being also able to travel over long distances [12]. Ultrafine particles (<100 nm diameter), also named nanoparticles, are produced through condensation of vapour molecules and form nuclei, instable particles that persist in atmosphere only for a short time (few minutes till few hours), close to their source, with rapid changes in number distribution. The inhaled NPs can pass through the respiratory membrane, entering the blood and travelling through the circulatory and lymphatic systems to the body's organs [13].

The atmospheric particle concentration is influenced by meteorological conditions. Wind speed improves the atmospheric air, especially if it is higher than 6 m/s. The temperature influences the air quality: a very high or a very low temperature is correlated with increased concentration of atmospheric particles because usually, these extreme temperatures occur synchronous with high atmospheric pressure that blocks the air motion. The high cloudiness is usually accompanied by low atmospheric pressure and high-speed winds, conditions that reduce the particles concentration in the atmosphere. The association of thin planetary boundary layer (peplosphere, the lower part of the troposphere) with high atmospheric pressure can block the particle motions, leading to an increased density of noxious substances in the air [14]. The urban areas located in valleys present atmosphere with reduced particle dispersion and have winds of low speed, conditions that lead to increased concentration of particles [15]. In the urban zones located in arid areas, the strong wind increases the particulate matter concentration [16].

2.1 Sources, chemical composition and morphological aspects of atmospheric nanoparticles

The diversity of atmospheric nanoparticles, as a result of highly variable sources and chemical composition, can explain the different effects on human body. Inorganic compounds (nitrates, ammonium, sulphates, trace metals) and organic compounds (hopanes from engine oils, organic acids) can attach noxious substances and can lead to severe diseases. The industrial zones release the atmosphere nanoparticles with high quantities of Fe, K (from biomass burning), or Ca (found in oil as additive) [17–19].

In urban atmosphere, different types of metallic NPs were identified, from which 50% contain two or more metals in their composition [20]. The most common metals are Fe, Ca, Al, Mg, Zn, Na, and they have many sources, from vehicles emission to industrial processes [21]. The principal source for metallic nanoparticles is represented by the vehicles through lubricating oil additives [22], diesel fuel additives [23], brake mechanism (releases 26–44% $PM_{0.1}$ because of metal volatilisation by heating) [24] and tyre dust (emits NPs of 15–50 nm diameter) [25]. Metallic NPs can also be produced through the wood combustion [26] and in port areas, by marine diesel engines [27]. Nanoparticles emitted by diesel engines contain a core made from solid elemental carbon and a layer of volatile organic carbon. They are of 10–50 nm diameter and can attach water-soluble ions, traces of elements from lubricants, or from engine functional abrasion [28].

Decamethylcyclopentasiloxane (D5), a volatile cyclic siloxane, found in personal care products is released in atmosphere in big quantities, with high concentrations during winter (~1 ng/m^3) and low concentrations during summer (~0.3 ng/m^3). D5 is transported in the atmosphere over long distances, but with seasonal variation, depending on the availability of OH radical ($OH^{\cdot}$) in the air, an oxygen species with which it reacts very quickly. Even if D5 can persist for almost 10 days in the atmosphere, it reaches the ground very rarely [29].

Silicon, a common compound of atmospheric nanoparticles, is detected in high concentrations in geographical areas with anthropogenic sources, with maximum levels during the daytime because Si-NPs production occurs through photochemical reactions. The oxidation of D5 leads to the development of Si nanoparticles in urban areas [30].

Morphological aspects of atmospheric nanoparticles differ in concordance with the emitting sources. The NPs identified in rail subways have specific chemical composition and specific morphology. Fe-NPs have irregular aspects: flake-like, botryoidal, crystal-like, or aggregated particles and most of them contain carbon. Calcium and silicon are often identified in the platform environment. In the subway's air, there also exist high concentrations of other metals (Zn, Ti, Sb) and traces of Al, K, Na, Mg, Cr, Co, S and Cl. Most of the inhaled Fe-NPs have the flake aspect, with different incorporated compounds [31].

2.2 Production of urban atmospheric nanoparticles

Ultrafine particles are produced through nucleation (vapour substances condense around the particles). An increase in atmospheric concentration of sulphuric acid leads, within 1–2 h, to an increase in particle number [32], with a reduced particle growth rate [33]. Sulphuric acid is considered an important factor for aerosol nucleation that leads to production of 2–3 nm NPs. Organic vapours can also produce nanoparticles with dimension that depend on the particle acidity, relative humidity and mechanisms of synthesis. Amines can lead to 4 nm NPs production, while organic carbonyls (aldehydes, α-dicarbonyls) can develop 4–6 nm particles [34].

Atmospheric nanoparticles can also be produced through evaporation of volatile compounds of larger particles [35].

The ultrafine particles with sizes between 1.5 and 2 nm form new larger particles through homogenous nucleation, process that occurs rapidly and leads to NPs dilution near the emission source [36]. Dry or wet depositions of particulate matter to the surfaces are important only for particles larger than 100 nm [37]. Most of the atmospheric NPs are too small to form cloud droplets that can be eliminated from the air by rain. The rainfall rate and the rain duration are important factors in NPs removing from the air, by production of larger particles that can be wet-deposited on surfaces [38].

There exist big differences in nanoparticles concentration measured in rural areas (2.6×10^3 to 4.8×10^3 particles/cm^3), in urban environments (42.1×10^3 to 48.2×10^3 particles/cm^3) [39] and in road tunnels (167.7×10^3 particles/cm^3) [40]. The rate of nanoparticles production is ~10^2 NPs/cm^3/s in urban areas and 10^4 to 10^5 NPs/cm^3/second in industrial and costal zones [41].

In urban atmosphere, the anthropogenic emissions represent the major sources of NPs, the vehicle emissions producing ~86% of the total atmospheric ultrafine particles [42]. There are many factors that can affect the NPs number concentration after vehicle emissions and among them, dilution is the most important event that depends on traffic conditions and has a duration of about 1 s [43]. Petrol (gasoline)-fuelled vehicles emit a low number of nanoparticles with sizes between 20 and 60 nm, while diesel-fuelled vehicles emit the most of atmospheric NPs but with a higher diameter, 20–130 nm [44]. An important factor for NPs emitted in the atmosphere is the driving manner: the petrol-fuelled vehicles driven at high speed (~120 km/h) and acceleration of these cars lead to emission of NPs in a similar number with the NPs number recorded at the diesel-fuelled cars emission [45]. The particles emitted by the vehicles are classified, according to their formation, into two groups: primary and secondary. The primary particles are released directly in the atmosphere like adsorbed or condensed hydrocarbons, sulphur compounds or metallic ash and their size is between 30 and 500 nm. The secondary particles are produced in the atmosphere from emitted hot gases that cool and condensate as nanoparticles with diameter lower than 30 nm, consisting of hydrated sulphuric acid and hydrocarbons [46].

Another factor that increases the particles concentration in the atmosphere is represented by the interaction between road and tyres [47].

In urban atmosphere, the industrial sources also contribute to NPs production, but only with 2%.

Nanotechnology developed new types of nanoparticles that are different in comparison with those found in the atmosphere, and even if these engineered NPs are incorporated in different products, they can escape in the environment, increasing the concentration of atmospheric ultrafine particles [48].

Cosmic dust enters the Earth atmosphere with 40,000 tons of particles/year. The cosmic particulate matter with low-velocity, mostly from asteroids and comets, can reach the ground [49, 50]. In the atmosphere, most of the cosmic particles are destroyed but some of $PM_{0.1}$ are vaporised and then recondensed into individual particles, being a source of iron nanoparticles [51].

2.3 Atmospheric nanoparticles in different urban areas

In urban atmosphere, the complex and turbulent mixing mechanisms influence the NPs flow around or over the buildings and streets.

The highest concentration of particulate matter is recorded in traffic intersections, but it is also increased in street canyons and on the roadsides, in comparison with the urban peripheral areas [52].

Within the street canyons, the nanoparticles' concentration depends on: traffic, atmospheric conditions (wind flow velocity, air temperature) [53, 54], street dimensions (large or narrow), street style (trees, street squares) and buildings' type (small or tall, balconies, walls roughness) [55].

In traffic intersection atmosphere, the concentration of nanoparticles emitted by cars depends on driving conditions. It is maximum during deceleration, it presents increased values during acceleration or cruising, and it is minimum during idling. The stop-and-go driving at the roads intersection produces NPs that can be identified in an area between 120 and 379 m, depending on intersection type [56].

The road tunnels environment has specific characteristics due to the "piston effect" produced by the traffic direction that generates a turbulent flow of mixed NPs-vehicle generated [57]. These specific properties of road tunnel air (that are not influenced by meteorological conditions), and the presence of high concentrations of precursor molecules provide the appropriate environment for nanoparticles formation (through nucleation) that are then transformed into larger molecules (through coagulation) [58].

In underground car park, the particulate matter concentration is higher than in opened urban area, with values that vary according to the traffic and dust presence on the ground. The entrance air in the park contains lesser concentrations of particulate matter than the exit air (acceleration, disc brake friction, dust re-suspension). The elements identified in underground car park environment are: Fe and Mn (the most abundant and related to dust re-suspension) and Zn and Cu (related to brake abrasion and tyre wear) [59].

On rail subways, the nanoparticles are made up of Fe, and their concentration, number and size are dependent on the train speed, pressure exerted on the rails, material of rails and wheels, lubricants used to reduce the wear, and on the temperature generated by the rail-wheel contact [60]. The nanoparticles concentrations are much higher than those above the ground [61]. Most of the NPs consist of Fe structures that are unique to the subways air, and their dimensions are very small [62].

3. Urban atmospheric nanoparticles in the human body: penetration, storage and elimination

The short-term exposure to atmospheric NPs exacerbates the chronic pulmonary and cardiovascular diseases and the long-term or repeated exposure can lead to death [63].

The respiratory system, digestive tract and skin are passed by the atmospheric nanoparticles [64], process that leads to oxidative stress, inflammation, or other pathological effects. The noxious effects of NPs depend on their size, chemical structure, or shape [65]. The skin penetration of atmospheric nanoparticles is realised transcellular (NPs <75 nm), intercellular, through the hair follicles, sweat and sebaceous glands [66]. Inside the dermal fibroblasts, nanoparticles generate genotoxic effects [67] and in the keratinocytes produce oxidative stress with DNA damages [68]. The structure of hair cuticula and of follicular canal permits mostly the passage of fine particles (~300–600 nm) [69]. The respiratory system is another entering route for nanoparticles through inhalation or through the wall of air passageways [70].

The storage of atmospheric nanoparticles is realised in many organs: lymph nodes, spleen, lungs, liver, brain, bladder, cardiovascular system, bone marrow, etc. [71].

The penetrated nanoparticles can be eliminated from the body through the faeces and urine, in quantities that are related to the NPs properties [72].

4. Oxidative stress produced by urban atmospheric nanoparticles

Space weather affects the human body, initiating different pathological mechanisms [73]. High-energy cosmic radiation may interact with atoms, compresses them, and pushes the electrons into the nucleus, leading to the transformation of protons into neutrons, with electron neutrinos release [74]. In the body's tissues with high amount of H^+, neutrons can produce the recoil protons, subatomic particles that can destroy the cells [75]. Neutrons can be carried inside the human body by the atmospheric NPs [76], where they can produce oxidative stress that damages the tissues.

Decamethylcyclopentasiloxane (D5), a highly volatile substance, affects the human body only after long-term exposures, leading to the decreased secretion of prolactin from the pituitary gland because of its indirect dopamine-like effects, or to the lung inflammation [77]. The respiratory system is affected by most of the air nanoparticles, the oxidative stress and sensitisation being the common mechanisms that initiate or exacerbate the asthma [78]. Oxidative stress promotes DNA alterations and other cellular disturbances that may lead to lung cancer [79].

The nanoparticles emitted by diesel engines have chemical structures (elemental carbon with absorbed organic compounds, metals, nitrates, sulphate, etc.) that can cause inflammation and oxidative stress in the lungs [80] but they can also pass into the brain [81] and initiate pathological processes in the central nervous system [82]. Inhaled diesel NPs not only affect the CNS through direct mechanisms because they can cross the blood-brain barrier, but also through indirect processes that involve the cytokines released from the areas where respiratory and cardiovascular inflammation develop. Microglia activation, oxidative stress, vessels microlesions and neurotoxic effect of cytokines stimulate the neurons degeneration [83]. Organic matter combustion or frictional heating of brake pads release in urban atmosphere nanoparticles of magnetite a ferromagnetic compound that contains Fe^{2+} and Fe^{3+} [84, 85]. They can enter the human brain through olfactory nerve and produce oxidative stress that can initiate neurodegeneration, process responsible for Alzheimer disease [86]. Al-NPs cause oxidative stress in brain vessels' endothelial cells and alter the expression of tight-junction proteins [87]. Nanoparticles can deposit in other parts of the brain (hypothalamus, cerebellum, frontal cortex, brain stem, etc.) because of their structural characteristics that permit the passage at the synaptic junctions and through blood-brain barrier [88], leading to neuroinflammation, similar to that identified in Alzheimer disease [89]. Diesel NPs can pass rapidly into the circulatory system. They produce endothelial dysfunction at the microcirculation level, accumulate at inflammation sites, especially in atherosclerotic plaques [90], and are able to pass into the foetal circulation through placenta barrier, being considered teratogen factors [91]. Inhalation of diesel-emitted NPs stimulates the platelet aggregation [92] and may trigger thrombogenesis. They stimulate lipid peroxidation in the tissues (small intestine, liver, etc.) and in the plasma, the oxidation of HDL altering the protective role of this lipoprotein [93]. The inflammation, increased oxidative stress [94], vasoconstriction, or endothelium dysfunctions promoted by the NPs prolonged exposure can lead to atherosclerosis [95]. The exposure to atmospheric NPs initiates the mechanisms that lead to myocardial ischemia in patients with coronary diseases [96]. Atmospheric $PM_{0.1}$ affects the blood flow into the vessels, decreasing the blood pressure [97] or increasing it [98], depending on the particulate matter and on the patient status. Nanoparticles can pass inside the cells and alter the functions of cellular proteins, cellular organelles and DNA [99]. Diesel nanoparticles damage the DNA through different mechanisms: direct attack on DNA, DNA oxidation, or inflammatory processes that generate excessive ROS with genotoxic effects [100]. The nanoparticles emitted by diesel engines or by other pollutant sources are able to initiate protein citrullination that leads to autoimmune diseases like collagen-induced arthritis [101].

Among the urban atmosphere nanoparticles, many are photochemically active (SiO_2, ZnO, TiO_2) and can produce exited electrons that can be transferred to the oxygen molecule, leading to superoxide radical ($O_2^{\cdot -}$) [102]. Most of the NPs contain Fe, Ca, Si, Zn, Cr, components that trigger the oxidative stress through Fenton or Haber-Weiss-type reactions, with production of $O_2^{\cdot -}$, $OH^{\cdot}$ and $1O_2^{\cdot}$ (singlet oxygen) [103]. NPs can alter the function of proteins [104] and lipids. Ti-NPs have affinity for lipids, being accumulated in cellular membrane in big quantities that rupture the membrane. Inside the cell membrane, Ti-NPs are covered by lipids, proteins and tissue factor and move away from the initial location. Internalisation of the covered Ti-NPs stimulates the lysosomes enzymes to degrade the nanoparticles' coat, letting the nanoparticles' active surface to attack the lysosomal membrane, leading in the end to the death of the cell [105]. Like Ti-NPs, also Zn-NPs and Ce-NPs can be stored in the cell membrane or in the organelles and can initiate oxidative stress [106].

The presence of Fe-NPs, especially in the subway environment, stimulates the oxidative stress and the DNA lesions [107]. Fe-NPs penetrate easily inside the cells, target the endoplasmic reticulum and mitochondria and lead to ROS synthesis [108]. Ca-NPs are transported rapidly into the circulatory system and can bind plasma proteins [109]. Ultrafine particles can pass easily into the cell increasing the cellular calcium concentration that stimulates the release of cytokines [110], inflammatory molecules that stimulate the oxidative stress [111].

The same effects of NPs were seen on erythrocytes: easy penetration of ultrafine particles, increased calcium concentration inside the cytoplasm, oxidative stress, processes that lead to haemolysis [112]. $PM_{0.1}$ exposure has effects also on white blood cells, producing an increase of monocytes concentration [113].

5. Conclusions

Urban atmospheric nanoparticles penetrate the human body and stimulate the oxidative stress in many organs leading to acute or chronic diseases, according to the time of exposure. The smaller nanoparticles penetrate easier inside the tissue cells, and their effects depend on the chemical composition, dimension, morphology, and on the reactive sites that are present on their surface. Urban atmosphere contains a complex mixture of different nanoparticles that are characteristic to specific urban locations. Prolonged exposure to high concentration of NPs in urban or industrial areas can lead to severe diseases or even death. New researches are required to complete the general view of the complex urban atmosphere.

Conflict of interest

The authors declare that there is no conflict of interest regarding the publication of this chapter.

Abbreviations

ROS	reactive oxygen species
NPs	nanoparticles; particles less than 100 nm in diameter
$PM_{0.1}$	particulate matter less than 0.1 μm in diameter
D5	decamethylcyclopentasiloxane
CNS	central nervous system
HDL	high-density lipoprotein

Author details

Daniela-Rodica Mitrea*, Alina-Mihaela Toader and Oana-Alina Hoteiuc
"Iuliu Hatieganu" University of Medicine and Pharmacy, Cluj-Napoca, Romania

*Address all correspondence to: rdmitrea@yahoo.co.uk

References

[1] Mohnen V, Hidy GM. Measurements of atmospheric nanoparticles (1875-1980). BAMS. 2010;**91**(11):1525-1539. DOI: 10.1175/2010BAMS2929.1

[2] Nel A, Xia T, Madler L, Li N. Toxic potential of materials at the nanolevel. Science. 2006;**311**(5761):622-627. DOI: 10.1126/science.1114397

[3] Clarke JC, Owens SR, Zhou J. An ultrafine sea-salt flux from breaking waves: Implications for cloud condensation nuclei in the remote marine atmosphere. Journal of Geophysical Research D: Atmospheres. 2006;**111**(D6):1-2. DOI: 10.1029/2005JD006565

[4] Zook HA. Spacecraft measurements of the cosmic dust flux. In: Peucker-Ehrenbrink B, Schmitz B, editors. Accretion of Extraterrestrial Matter Throughout Earth's History. Boston: Springer; 2001. pp. 75-92. DOI: 10.1007/978-1-4419-8694-8

[5] Holmes NS. A review of particle formation events and growth in the atmosphere in the various environments and discussion of mechanistic implications. Atmospheric Environment. 2007;**25**(Suppl. 1):2183-2201. DOI: 10.1016/j.atmosenv.2006.10.058

[6] Tunved P, Hansson H, Kerminen V, Strom J, Dal Maso M, Lihavainen H, et al. High natural aerosol loading over boreal forests. Science. 2006;**312**(5771):261-263. DOI: 10.1126/science.1123052

[7] Oberdörster G, Oberdörster E, Oberdörster J. Nanotoxicology: An emerging discipline evolving from studies of ultrafine particles. Environmental Health Perspectives. 2005;**113**(7):823-839. DOI: 10.1289/ehp.7339

[8] Strambeanu N, Demetrovici L, Dragos D. Anthropogenic sources of nanoparticles. In: Lungu M, Neculae A, Bunoiu M, Biris C, editors. Nanoparticles' Promises and Risks: Characterization, Manipulation, and Potential Hazards to Humanity and the Environment. Switzerland: Springer International Publishing; 2015. pp. 21-54. DOI: 10.1007/978-3-319-11728-7

[9] Johnston MV, Klems JP, Zordan CA, Pennington MR, Smith JN, HEI Health Review Committee. Selective detection and characterisation of nanoparticles from motor vehicles. Research Report. Health Effects Institute. 2013;**173**:3-45

[10] Slezakova K, Morais S, Pereira MC. Atmospheric nanoparticles and their impacts on public health. In: Rodriguez-Morales AJ, editor. Current Topics in Public Health. Rijeka, Croatia: IntechOpen; 2013. pp. 503-529. DOI: 10.5772/54775

[11] Kumar P, Fennell BR. Measurements of particles in the 5-1000 nm range close to road level in an urban street canyon. Science of the Total Environment. 2008;**390**(2-3):437-447. DOI: 10.1016/j.scitotenv.2007.10.013

[12] Anastasio C, Martin ST. Atmospheric nanoparticles. Reviews in Mineralogy and Geochemistry. 2001;**44**(1):293-349. DOI: 10.2138/rmg.2001.44.08

[13] Oberdörster G, Sharp Z, Atudorei V, Elder A, Gelein R, Kreyling W, et al. Translocation of inhaled ultrafine particles to the brain. Inhalation Toxicology. 2004;**16**(6-7):437-445. DOI: 10.1080/08958370490439597

[14] Jędruszkiewicz J, Czernecki B, Marosz M. The variability of PM10 and PM2.5 concentrations in selected Polish agglomerations: The role of meteorological conditions, 2006-2016. IJEHR. 2017;**27**(6):441-462. DOI: 10.1080/09603123.2017.1379055

[15] Giri D, Krishna Murthy V, Adhikary PR. The influence of meteorological conditions on PM10 concentrations in Kathmandu valley. International Journal of Environmental Research. 2008;**2**(1):49-60

[16] Al Jallad F, Al Katheri E, Al Omar M. Concentrations of particulate matter and their relationships with meteorological variables. Sustainable Environment Research. 2013;**23**(3):191-198

[17] Seinfeld JH, Pandis SN, editors. Atmospheric Chemistry and Physics: From Air Pollution to Climate Change. 2nd ed. Hoboken, New Jersey: John Wiley & Sons; 2006. pp. 19-55. ISBN 978-0-471-72018-8

[18] Terzano C, Di Stefano F, Conti V, Graziani E, Petroianni A. Air pollution ultrafine particles: Toxicity beyond the lung. European Review for Medical and Pharmacological Sciences. 2010;**14**(10):809-821

[19] Chow JC, Watson JG. Review of measurement methods and composition of ultrafine particles. Aerosol and Air Quality Research. 2007;7(2):121-173

[20] Adachi K, Buseck PR. Hosted and free-floating metal-bearing atmospheric nanoparticles in Mexico City. Environmental Science & Technology. 2010;**44**:2299-2304. DOI: 10.1021/es902505b

[21] Sanderson P, Delgado-Saborit JM, Harrison RM. A review of chemical and physical characterisation of atmospheric. Atmospheric Environment. 2014;**94**:353-365. DOI: 10.1016/j.atmosenv.2014.05.023

[22] Miller AL, Stipe CB, Habjan MC, Ahlstrand GG. Role of lubrication oil in particulate emissions from a hydrogen-powered internal combustion engine. Environmental Science & Technology. 2007;**41**:6828-6835. DOI: 10.1021/es070999r

[23] Cassee FR, Van Balen EC, Singh C, Green D, Muijser H, Weinstein J, et al. Exposure, health and ecological effects review of engineered nanoscale cerium and cerium oxide associated with its use as a fuel additive. Critical Reviews in Toxicology. 2011;**41**:213-229. DOI: 10.3109/10408444.2010.529105

[24] Garg BD, Cadle SH, Mulawa PA, Groblicki PJ, Laroo C, Parr GA. Brake wear particulate matter emissions. Environmental Science & Technology. 2000;**34**:4463-4469. DOI: 10.1021/es001108h

[25] Dahl A, Gharibi A, Swietlicki E, Gudmundsson A, Bohgard M, Ljungman A, et al. Traffic-generated emissions of ultrafine particles from pavement-tire interface. Atmospheric Environment. 2006;**40**:1314-1323. DOI: 10.1016/j.atmosenv.2005.10.029

[26] Tissari J, Lyyränen J, Hytönen K, Sippula O, Tapper U, Frey A, et al. Fine particle and gaseous emissions from normal and smouldering wood combustion in a conventional masonry heater. Atmospheric Environment. 2008;**42**:7862-7873. DOI: 10.1016/j.atmosenv.2008.07.019

[27] Lyyränen J, Jokiniemi J, Kauppinen EI, Joutsensaari J. Aerosol characterisation in medium-speed diesel engines operating with heavy fuel oils. Journal of Aerosol Science. 1999;**30**:771-784. DOI: 10.1016/S0021-8502(98)00763-0

[28] Alander TJ, Leskinen AP, Raunemaa TM, Rantanen L. Characterization of diesel particles: Effects of fuel reformulation, exhaust aftertreatment, and engine operation on particle carbon composition and volatility. Environmental Science & Technology. 2004;**38**(9):2707-2714. DOI: 10.1021/es030129j

[29] McLachlan MS, Kierkegaard A, Hansen KM, Van Egmond R, Christensen JH, Skjøth CA. Concentration and fate of decamethylcyclopentasiloxane (D5) in the Atmposphere. Environmental Science & Technology. 2010;**44**:5365-5370. DOI: 10.1021/es100411w

[30] Bzdek BR, Horan AJ, Pennigton MR, Janechek NJ, Baek J, Stanier CO, et al. Silicon is a frequent component of atmospheric nanoparticles. Environmental Science & Technology. 2014;**48**:11137-11145. DOI: 10.1021/es5026933

[31] Moreno T, Martins V, Querol X, Jones T, BéruBé K, Minguillón MC, et al. A new look at inhalable metalliferous airborne particles on rail subway platforms. Science of the Total Environment. 2015;**505**:367-375. DOI: 10.1016/j.scitotenv.2014.10.013

[32] Weber R, Marti JJ, McMurray P, Eisele FL, Tanner DJ, Jefferson A. Measured atmospheric new particle formation rates: Implications for nucleation mechanisms. Chemical Engeneering Communications. 1996;**151**(1):53-64. DOI: 10.1080/00986449608936541

[33] Weber R, Marti JJ, McMurray P, Eisele FL, Tanner DJ, Jefferson A. Measurement of new particle formation and ultrafine particle growth rates at a clean continental site. Journal of Geophysical Research D: Atmospheres. 1997;**102**(4):4375-4386. DOI: 10.1029/96JD03656

[34] Wang L, Khalizov AF, Zheng J, Xu W, Ma Y, Lal V, et al. Atmospheric nanoparticles formed from heterogenous reaction of organics. Nature Geoscience. 2010;**3**:238-242. DOI: 10.1038/NGEO778

[35] Zhang KM, Wexler AS. Evolution of particle number distribution near roadways—Part I: Analysis of aerosol dynamics and its implications for engine emission measurement. Atmospheric Environment. 2004;**38**:6643-6653. DOI: 10.1016/j.atmosenv.2004.06.043

[36] Wehner B, Birmili W, Gnauk T, Wiedensohler A. Particle number size distributions in a street canyon and their transformation into the urban air background: Measurements and a simple model study. Atmospheric Environment. 2002;**36**(13):2215-2223. DOI: 10.1016/S1352-2310(02)00174-7

[37] Petroff A, Zhang L. Development and validation of a size-resolved particle dry deposition scheme for applications in aerosol transport models. Geoscientific Model Development. 2010;**3**(2):753-769. DOI: 10.5194/gmd-3-753-2010

[38] Andronache C. Precipitation removal of ultrafine aerosol particles from the atmospheric boundary layer. Journal of Geophysical Research Atmospheres. 2004;**109**:D16S07. DOI: 10.1029/2003JD004050

[39] Morawska L, Ristovski Z, Jayaratne ER, Keogh DU, Ling X. Ambient nano and ultrafine particles from motor vehicle emissions: Characteristics, ambient processing and implications on human exposure. Atmospheric Environment. 2008;**42**(35):8113-8138. DOI: 10.1016/j.atmosenv.2008.07.050

[40] Van Dingenen R, Raes F, Putaud J-P, Baltensperger U, Charron A, Facchini M-C, et al. A European aerosol phenomenology—1: Physical characteristics of particulate matter at kerbside, urban, rural and background sites in Europe. Atmospheric Environment. 2004;**38**(16):2561-2577. DOI: 10.1016/j.atmosenv.2004.01.040

[41] Kulmala M, Vehkamäki H, Petäjä T, Dal Maso M, Lauri A, Kerminen V-M,

et al. Formation and growth rates of ultrafine atmospheric particles: A review of observations. Journal of Aerosol Science. 2004;**35**:143-176. DOI: 10.1016/j.jaerosci.2003.10.003

[42] Pey J, Querol X, Alastuey A, Rodriguez S, Putaud JP, Van Dingenen R. Source apportionment of urban fine and ultra fine particle number concentration in a Western Mediterranean City. Atmospheric Environment. 2009;**43**(29):4407-4415. DOI: 10.1016/j.atmosenv.2009.05.024

[43] Kumar P, Ketzel M, Vardoulakis S, Pirjola L, Britter R. Dynamics and dispersion modelling of nanoparticles from road traffic in the urban atmospheric environment—A review. Journal of Aerosol Science. 2011;**42**:580-603. DOI: 10.1016/j.jaerosci.2011.06.001

[44] Harris SJ, Maricq MM. Signature size distribution for diesel and gasoline engine exhaust particulate matter. Journal of Aerosol Science. 2001;**32**(6):749-764. DOI: 10.1016/S0021-8502(00)00111-7

[45] Graskow BR, Kittelson DB, Abdul-Khaleek IS, Ahmadi MR, Morris JE. Characterization of exhaust particulate emission from a spark ignition engine. SAE Special Publications. 1998;**1326**:155-165. DOI: 10.4271/980528

[46] Kittelson DB, Watts WF, Johnson JP. Nanoparticle emission on Minnesota highways. Atmospheric Environment. 2004;**38**(1):9-19. DOI: 10.1016/j.atmosenv.2003.09.037

[47] Gustafsson M, Blomqvist G, Gudmundsson A, Dahl A, Swietlicki E, Boghard M, et al. Properties and toxicological effects of particles from the interaction between tyres, road pavement and winter traction material. Science of the Total Environment. 2008;**393**(2-3):226-240. DOI: 10.1016/j.scitotenv.2007.12.030

[48] Bystrzejewska-Piotrowska G, Golimowski J, Urban PL. Nanoparticles: Their potential toxicity, waste and environmental management. Waste Management. 2009;**29**(9):2587-2595. DOI: 10.1016/j.wasman.2009.04.001

[49] Genge MJ. Koronis asteroid dust within Antarctic ice. Geology. 2008;**36**(9):687-690. DOI: 10.1130/G24493A.1

[50] Genge MJ, Larsen J, Van Ginneken M, Suttle MD. An urban collection of modern-day large micrometeorites: Evidence for variations in the extraterrestrial dust flux through the quaternary. Geology. 2017;**45**(2):119-122. DOI: 10.1130/G38352.1

[51] Masciangioli T, Alper J. In: A Workshop Summary National Research Council, editor. Challenges in Characterizing Small Particles. Exploring Particles from the Nano- to Microscale. Washington DC: The National Academies Press; 2012. DOI: 10.17226/13317

[52] Schneider IL, Teixeira EC, Oliveira LFS, Wiegand F. Atmospheric particle number concentration and size distribution in a traffic-impacted area. Atmospheric Pollution Research. 2015;**6**:877-885. DOI: 10.5094/APR.2015.097

[53] Solazzo E, Cai X, Vardoulakis S. Modelling wind flow and vehicle-induced turbulence in urban streets. Atmospheric Environment. 2008;**42**:4918-4931. DOI: 10.1016%2Fj.atmosenv.2008.02.032

[54] Xie X, Huang Z, Wang J, Xie Z. Thermal effects on vehicle emission dispersion in an urban street canyon. Transportation Research Part D. Transport and Environment. 2005;**10**:197-212. DOI: 10.1016/j.trd.2005.01.002

[55] Gayev YA, Savory E. Influence of street obstructions on flow processes

within urban canyons. Journal of Wind Engeneering and Industrial Aerodynamics. 1999;**82**:89-103. DOI: 10.1016/S0167-6105(98)00212-8

[56] Goel A, Kumar P. Zone of influence for particle number concentrations at signalised traffic intersections. Atmospheric Environment. 2015;**123**:25-38. DOI: 10.1016/j.atmosenv.2015.10.054

[57] Bari S, Naser J. Simulation of air flow and pollution levels caused by severe traffic jam in a road tunnel. Tunnelling and Underground Space Technology. 2010;**25**(1):70-77. DOI: 10.1016/j.tust.2009.09.004

[58] Cheng Y-H, Liu Z-S, Chen C-C. On-road measurements of ultrafine particle concentration profiles and their size distributions inside the longest highway tunnel in Southeast Asia. Atmospheric Environment. 2010;**44**(6):763-772. DOI: 10.1016/j.atmosenv.2009.11.040

[59] Li Y, Xiang R. Particulate pollution in an underground car park in Wuhan, China. Particuology. 2013;**11**:94-98. DOI: 10.1016/j.partic.2012.06.010

[60] Lee Y, Choi K, Jung W, Versoza ME, Barabad MLM, Kim T, et al. Generation characteristics of nanoparticles emitted from subways in operation. Aerosol and Air Quality Research. 2018;**18**:2230-2239. DOI: 10.4209/aaqr.2017.11.0439

[61] Moreno T, Pérez N, Reche C, Martins V, de Miguel E, Capdevila M, et al. Subway platform air quality: Assessing the influences of tunnel ventilation, train piston effect and station design. Atmospheric Environment. 2014;**92**:461-468. DOI: 10.1016/j.atmosenv.2014.04.043

[62] Loxham M, Cooper MJ, Gerlofs-Nijland ME, Cassee FR, Davies DE, Palmer MR, et al. Physicochemical characterization of airborne particulate matter at a mainline underground railway station. Environmental Science & Technology. 2013;**47**(8):3614-3622. DOI: 10.1021/es304481m

[63] Brugge D, Durant JL, Rioux C. Near-highway pollutants in motor vehicle exhaust: A review of epidemiologic evidence of cardiac and pulmonary health risks. Environmental Health. 2007;**6**:23. DOI: 10.1186/1476-069X-6-23

[64] Helland A, Kastenholtz H, Thidell A, Arnfalk P, Deppert K. Nanoparticulate material and regulatory policy in Europe: An analysis of stakeholder perspectives. Journal of Nanoparticle Research. 2006;**8**(5):709-719. DOI: 10.1007/s11051-006-9096-3

[65] Clichici S, Filip A. In vivo assessment of Nanomaterials toxicity. In: Larramendy LM, Soloneski S, editors. Nanomaterials—Toxicity and Risk Assessment. Rijeka, Croatia: IntechOpen; 2015. pp. 93-121. DOI: 10.5772/60707

[66] Rancan F, Gao Q, Graf C, Troppens S, Hadam S, Hackbarth S, et al. Skin penetration and cellular uptake of amorphous silica nanoparticles with variable size, surface functionalization, and colloidal stability. ACS Nano. 2012;**6**(8):6829-6842. DOI: 10.1021/nn301622h

[67] Quignard S, Mosser G, Boissiere M, Coradin T. Long-term fate of silica nanoparticles interacting with human dermal fibroblasts. Biomaterials. 2012;**33**:4431-4442. DOI: 10.1016/j.biomaterials.2012.03.004

[68] Nabeshi H, Yoshikawa T, Matsuyama K, Nakazato Y, Tochigi S, Kondoh S, et al. Amorphous nanosilica induce endocytosis dependent ROS generation and DNA damage in human keratinocytes. Particle and Fibre Toxicology. 2011;**8**(1):1-10. DOI: 10.1186/1743-8977-8-1

[69] Lademann JM, Patzelt A, Richter H, Antoniou C, Sterry W, Knorr F.

Determination of the cuticula thickness of human and porcine hairs and their potential influence on the penetration of nanoparticles into the hair follicles. Journal of Biomedical Optics. 2009;**14**(2):021014-1-021014-4. DOI: 10.1117/1.3078813

[70] Qiao H, Liu W, Gu H, Wang D, Wang Y. The transport and deposition of nanoparticles in respiratory system by inhalation. Journal of Nanomaterials. 2015;**2015**:1-8. DOI: 10.1155/2015/394507

[71] Nemmar A, Hoet PHM, Vanquickenborne B, Dinsdale D, Thomeer M, Hoyaerts MF, et al. Passage of inhaled particles into the blood circulation in humans. Circulation. 2002;**105**:411-414. DOI: 10.1161/hc0402.104118

[72] Illum L, Davis SS, Müller RH, Mak E, West P. The organ distribution and circulation time of intravenously injected colloidal carriers sterically stabilized with a blockcopolymer—Poloxamine 908. Life Science. 1987;**40**(4):367-374. DOI: 10.1016/0024-3205(87)90138-x

[73] Oranevski VN, Breus TK, Baevski RM, Rapoport SI, Petrov VM, Barsukova ZV. Effect of geomagnetic ctivity on the functional status of the body. Biofizika. 1998;**43**(5):819-826

[74] Segré E. Chapter 3: K-electron capture by nuclei. In: Trower WP, editor. Discovering Alvarez: Selected Works of Luis W. Alvarez, with Commentary by his Students and Colleagues. Chicago IL USA: University of Chicago Press; 1987. pp. 11-12. ISBN-10: 0226813045

[75] Mitrea DR, Mortazavi Moshkenani H, Hoteiuc OA, Bidian C, Toader AM, Clichici S. Antioxidant protection against cosmic radiation-induced oxidative stress at commercial flight altitude. Journal of Physiology and Pharmacology. 2018;**69**(4):1899-1905. DOI: 10.26402/jpp.2018.4.03

[76] Stoupel EG. Cosmic ray (neutron) activity and air pollution nanoparticles-cardiovascular disease risk factors-separate or together? Journal of Basic and Clinical Physiology and Pharmacology. 2016;**27**(5):493-496. DOI: 10.1515/jbcpp-2015-0119

[77] Dekant W, Klaunig JE. Toxicology of decamethylcyclopentasiloxane (D5). Regulatory Toxicology and Pharmacology. 2016;**74**(Suppl):S67-S76. DOI: 10.1016/j.yrtph.2015.06.011

[78] Guarnieri M, Balmes JR. Outdoor air pollution and asthma. Lancet. 2014;**383**(9928):1581-1592. DOI: 10.1016/S0140-6736(14)60617-6

[79] Lawless MW, O'Byrne KJ, Grayb SG. Oxidative stress induced lung cancer and COPD: Opportunities for epigenetic therapy. Journal of Cellular and Molecular Medicine. 2009;**13**(9a):2800-2821. DOI: 10.1111/j.1582-4934.2009.00845.x

[80] Greim H, Borm P, Schins R, Do K. Toxicity of fibres and particles. Report of the workshop held in Munich, Germany, 26-27 October 2000. Inhalation Toxicology. 2001;**13**(9):737-754. DOI: 10.1080/08958370118273

[81] Betzer O, Shilo M, Opochinsky R, Barnoy E, Motiei M, Okun E, et al. The effect of nanoparticle size on the ability to cross the blood–brain barrier: An in vivo study. Nanomedicine. 2017;**12**(13):1533-1546. DOI: 10.2217/nnm-2017-0022

[82] Elder A, Gelein R, Silva V, Feikert T, Opanashuk L, Carter J, et al. Translocation of inhaled ultrafine manganese oxide particles to the central nervous system. Environmental Health Perspectives. 2006;**114**(8):1172-1178. DOI: 10.1289/ehp.9030

[83] Wang Y, Tang M. Toxicity of diesel exhaust particles on central nervous system. Nanomedicine: NBM. 2018;**14**(5):1794. DOI: 10.1016/j.nano.2017.11.151

[84] McClean RG, Kean WF. Contributions of wood ash magnetism to archaeomagnetic properties of fire pits and hearths. Earth and Planetary Science Letters. 1993;**119**(3):387-394. DOI: 10.1016/0012-821X(93)90146-Z

[85] Kukutschová J, Moravec P, Tomášek V, Matějka V, Smolík J, Schwarz J, et al. On airborne nano/micro-sized wear particles released from low-metallic automotive brakes. Environmental Pollution. 2011;**159**(4):998-1006. DOI: 10.1016/j.envpol.2010.11.036

[86] Maher BA, Ahmed IAM, Karloukovski V, MacLaren DA, Foulds PG, Allsop D, et al. Magnetite pollution nanoparticles in the human brain. Proceedings of the National Academy of Science. 2016;**113**(39):10797-10801. DOI: 10.1073/pnas.1605941113

[87] Chen L, Yokel RA, Henning B, Toborek M. Manufactured aluminium oxide nanoparticles decrease expression of tight junction proteins in brain vasculature. Journal of Neuroimmune Pharmacology. 2008;**3**:286-295. DOI: 10.1007/s11481-008-9131-5

[88] Dorman DC, McManus BE, Parkinson CU, Manuel CA, McElveen AM, Everitt JI. Nasal toxicity of manganese sulfate and manganese phosphate in young male rats following subchronic (13-week) inhalation eposure. Inhalation Toxicology. 2004;**16**:481-488. DOI: 10.1080/08958370490439687

[89] Calderón-Garcidueñas L, Reed W, Maronpot RR, Henríquez-Roldán C, Delgado-Chavez R, Calderón-Garcidueñas A, et al. Brain inflammation and Alzheimer's-like pathology in individuals exposed to severe air pollution. Toxicologic Pathology. 2004;**32**:650-658. DOI: 10.1080/01926230490520232

[90] Miller MR, Raftis JB, Langrish JP, McLean SG, Samutrtai P, Connell SP, et al. Inhaled nanoparticles accumulate at sites of vascular disease. ACS Nano. 2017;**11**:4542-4552. DOI: 10.1021/acsnano.6b08551

[91] Keelan JA. Nanoparticles versus the placenta. Nature Nanotechnology. 2011;**6**(5):263-264. DOI: 10.1038/nnano.2011.65

[92] Rajagopalan S, Al-Kindi SG, Brook RD. Air pollution and cardiovascular disease. JACC state-of-the-art review. Journal of the American College of Cardiology. 2018;**72**(17):2054-2070. DOI: 10.1016/j.jacc.2018.07.099

[93] Yin F, Lawal A, Ricks J, Fox JR, Larson T, Navab M, et al. Diesel exhaust induces systemic lipid peroxidation and development of dysfunctional pro-oxidant and pro-inflammatory high-density lipoprotein. Arteriosclerosis, Thrombosis, and Vascular Biology. 2013;**33**:1153-1161. DOI: 10.1161/ATVBAHA.112.300552

[94] Araujo JA, Barajas B, Kleinman M, Wang X, Bennett BJ, Gong KW, et al. Ambient particulate pollutants in the ultrafine range promote early atherosclerosis and systemic oxidative stress. Circulation Research. 2008;**102**:589-596. DOI: 10.1161/CIRCRESAHA.107.164970

[95] Frampton MW, Stewart JC, Oberdörster G, Morrow PE, Chalupa D, Pietropaoli AP, et al. Inhalation of ultrafine particles alters blood leukocyte expression of adhesion molecules in humans. Environmental Health Perspectives. 2006;**114**(1):51-58. DOI: 10.1289/ehp.7962

[96] Pekkanen J, Peters A, Hoek G, Tiitanen P, Brunekreef B, De Hartog J, et al. Particulate air pollution and risk of ST-segment depression during repeated submaximal exercise tests among subjects with coronary heart disease: The exposure and risk assessment for fine and ultrafine particles in ambient air (ULTRA) study. Circulation. 2002;**106**(8):933-938. DOI: 10.1161/01.CIR.0000027561.41736.3C

[97] Ibald-Mulli A, Timonen KL, Peters A, Heinrich J, Wolke G, Lanki T, et al. Effects of particulate air pollution on blood pressure and heart rate in subjects with cardiovascular disease: A multicenter approach. Environmental Health Perspectives. 2004;**112**(3):369-377. DOI: 10.1289/ehp.6523

[98] Sun Q, Yue P, Ying Z, Cardounel AJ, Brook RD, Devlin R, et al. Air pollution exposure potentiates hypertension through reactive oxygen species mediated activation of rho/ROCK. Arteriosclerosis, Thrombosis, and Vascular Biology. 2008;**28**(10):1760-1766. DOI: 10.1161/ATVBAHA.108.166967

[99] Klein J. Probing the interactions of proteins and nanoparticles. Proceedings of the National Academy of Sciences of the United States of America. 2007;**104**:2029-2030. DOI: 10.1073/pnas.0611610104

[100] Dybdahl M, Risom L, Bornholdt J, Autrup H, Loft S, Wallin H. Inflammatory and genotoxic effects of diesel particles in vitro and in vivo. Mutation Research. 2004;**562**(1-2):119-131. DOI: 10.1016/j.mrgentox.2004.05.010

[101] Mohamed BM, Verma NK, Davies AM, McGowan A, Crosbie-Staunton K, Prina-Mello A, et al. Citrullination of proteins: A common post-translational modification pathway induced by different nanoparticles in vitro and in vivo. Nanomedicine. 2012;**7**(8):1181-1195. DOI: 10.2217/nnm.11.177

[102] Hoffmann M, Holtze EM, Wiesner MR. Reactive oxygen species generation on nanoparticulate material. In: Wiesner MR, Bottero JY, editors. Environmental Nanotechnology. Applications and Impacts of Nanomaterials. New York, USA: McGraw Hill; 2007. pp. 155-203. ISBN: 9780071477505

[103] Knaapen AM, Borm PJA, Albrecht C, Schins RPF. Inhaled particles and lung cancer, part A: Mechanisms. International Journal of Cancer. 2004;**109**(6):799-809. DOI: 10.1002/ijc.11708

[104] Saptarshi SR, Duschl A, Lopata AL. Interaction of nanoparticles with proteins: Relation to bio-reactivity of the nanoparticle. Journal of Nanobiotechnology. 2013;**11**(26):1-12. DOI: 10.1186/1477-3155-11-26

[105] Urbančič I, Garvas M, Kokot B, Majaron H, Umek P, Cassidy H, et al. Nanoparticles can wrap epithelial cell membranes and relocate them across the epithelial cell layer. Nano Letters. 2018;**18**(8):5294-5305. DOI: 10.1021/acs.nanolett.8b02291

[106] Buzea C, Pacheco I, Robbie K. Nanomaterials and nanoparticles: Sources and toxicity. Biointerphases. 2007;**2**(4):MR17-MR71. DOI: 10.1116/1.2815690

[107] Jung MH, Kim HR, Park YJ, Park DS, Chung KH, Oh SM. Genotoxic effects and oxidative stress induced by organic extracts of particulate matter (PM 10) collected from a subway tunnel in Seoul, Korea. Mutation Research. 2012;**749**(1-2):39-47. DOI: 10.1016/j.mrgentox.2012.08.002

[108] Park EJ, Umh HN, Choi DH, Cho MH, Choi W, Kim SW, et al. Magnetite- and maghemite-induced different toxicity in murine alveolar macrophage

cells. Archives of Toxicology. 2014;**88**(8):1607-1618. DOI: 10.1007/s00204-014-1210-1

[109] Lee JA, Kim MK, Kim HM, Lee JK, Jeong J, Kim YR, et al. The fate of calcium carbonate nanoparticles administered by oral route: Absorption and their interaction with biological matrices. International Journal of Nanomedicine. 2015;**10**:2272-2293. DOI: 10.2147/IJN.S79403

[110] Mitrea DR, Clichici S, Filip A, Olteanu D, Baldea I, Moldovan R, et al. Resveratrol and loratadine effects on oxidative stress induced by experimental inflammation. Studia UBB Chemia. 2017;**LXII**(1):89-100. DOI: 10.24193/subbchem.2017.1.07

[111] Yang D, Elner SG, Bian Z-M, Till GO, Petty HR, Elnera VM. Pro-inflammatory cytokines increase reactive oxygen species through mitochondria and NADPH oxidase in cultured RPE cells. Experimental Eye Research. 2007;**85**(4):462-472. DOI: 10.1016/j.exer.2007.06.013

[112] Nemmar A, Beegam S, Yuvaraju P, Yasin J, Shahin A, Ali BH. Interaction of amorphous silica nanoparticles with erythrocytes in vitro: Role of oxidative stress. Cellular Physiology and Biochemistry. 2014;**34**:255-265. DOI: 10.1159/000362996

[113] Malorni L, Langella MG, Iavicoli I, Pedata P. Role of ultrafine nanoparticles in lung cancer. Journal of Bioanalysis & Biomedicine. 2017;**9**(5):244-249. DOI: 10.4172/1948-593X.1000187

Chapter 5

The Interaction of Tungsten Dust with Human Skin Cells

Lavinia Gabriela Carpen, Tomy Acsente, Maria Adriana Acasandrei, Elena Matei, Claudia Gabriela Chilom, Diana Iulia Savu and Gheorghe Dinescu

Abstract

In this chapter, we evaluate the tungsten (W) nanoparticle toxicity with respect to the normal human skin fibroblast cell. Tungsten dust formation is expected in the tokamak-type nuclear fusion installations, regarded as future devices for large-scale, sustainable, and carbon-free energy. This dust, composed of tungsten particles of variable size, from nanometers to micrometers, could be harmful to humans in the case of loss of vacuum accident (LOVA). In order to undertake the toxicity studies, tokamak-relevant dust has been deliberately produced in laboratory and afterward analyzed. Following that, cytotoxicity tests were performed using normal human skin fibroblast cell lines, BJ ATCC CRL 2522. Our study concludes that, at a low concentration (until 100 μg/mL), no cytotoxic effect of tungsten nanoparticles was observed. In contrast, at higher concentrations (up to 2 mg/mL), nanometric dust presents toxic effects on the cells.

Keywords: toxicity, tungsten nanoparticles, plasma technology, magnetron sputtering, fusion technology

1. Introduction

Over the past decade, nanotechnology has received a lot of attention, due to its diversity of applications, ranging from electronics, aerospace, computers to biology and nanomedicine. All aspects of our lives have slowly begun to rely on this nanotechnology revolution. Nanotechnology includes the study and manipulation of nanoscale particles. Therefore, this revolution requires the large-scale production of these nanostructures [1]. Due to much smaller dimensions, this nanomaterial presents specific properties, sometimes in contrast with the bulk materials. Beside useful characteristics such as higher values of specific surface/volume ratio, enhanced chemical reactivity, mechanical or physicochemical properties as well as distinct optical properties relative to scale-up materials, these nanomaterials can present a danger to human health. The growing popularity of nanotechnology in medicine has been limited because of the potential side effects caused by the possible toxicity of the nanoparticles [2]. Adverse health effects can be caused by inhalation, oral ingestion, and absorption through the skin or by injection [3]. The studies indicated that the nanoparticle toxicity is observed through cellular modifications that include cell membrane damage, mitochondrial function affection, apoptosis activation, and also through increased oxidative mechanism stress [4]. These issues have underpinned the development of a new branch of research, under the name of nanotoxicology. Nanotoxicology includes the life cycle study of nanoparticles, in order to better understand the health risks involved in their use. Specific properties, such as particle size, nanoparticle shape, and surface/volume ratio, are considered important factors that directly contribute to nanomaterial toxicity. For example, carbon nanofibers, single-wall nanotubes, and multiwall nanotubes have shown that the toxicity of carbon material depends exclusively on the particles' shape and size [5]. In vivo experiments showed that carbon nanotubes cause granulomatous lung lesions [6] and can also induce platelet aggregation that highlighted a very toxic effect of carbon nanoparticles on the human body.

An element of major importance but not sufficiently introduced in in vitro toxicity evaluation is related to one of the largest organs of the body, the skin. Skin is the barrier between the internal organisms and the external environment. The skin is composed of multilayers and presents on the surface sweat pores, sebaceous glands, and hair sites necessary for hair growth [7]. This is the reason why this large exposure area (18,000 cm^2) serves as one of the main inlet ports of nanomaterials inside the body. The tendency of nanomaterials to pass through the skin, thus referring to the mechanism of skin absorption or penetration, is a major factor in the nanomaterial dermatotoxic potential in general, and tungsten (W) nanomaterials especially, mainly taking into account the purpose of this work. The dermatotoxicity produced by nanoparticles was reported primarily by the in vitro experiments. Bennat and Goymann [2, 8] have shown that nanoparticles can penetrate the surface of the skin much more easily by entering through the pores or hair follicles. In general, skin exposure to nanoparticles is mediated through the nanomaterials which are contained in cosmetics or wound dressings. For example, it has been found that sunscreens containing TiO_2 have seeped into the deep parts of the epidermis through the hair sites. Moreover, it's a known fact that silver nanoparticles are effective in treating patients with burns. For example, in the case of Acticoat, a dressing that is coated with silver nanoparticles that is often used for this purpose, there have been various studies that have reported the safety of this drug. Despite this, the toxicity of silver nanoparticles has been reported by a patient with burns of over 30% of the body surface, when they were treated with this silver-coated dressing [2, 9]. Subsequently, dermatotoxicity produced by nanoparticles was also reported in in vivo experiments. Using an animal model study, beryllium particles as large as 0.5 and 1.0 μm in diameter have been shown to be able to penetrate even the epidermis [10], producing toxicity increase on an animal skin.

1.1 Toxicology studies devoted to tungsten at micro- and nanoscale

A nanomaterial with various and multiple applications in the industry, which are rapidly starting to be investigated from a cytotoxicity point of view, is represented by metal nanoparticles. These nanoparticles are widely used in industry and biotechnology, and they are also of great importance in military applications [4]. Thus, the fact that these nanomaterials present potential harmful effects must be urgently researched, especially because they are to be used increasingly [11].

A particular case of metal material is represented by tungsten, a metal with exceptional physical and chemical characteristics. This hard-steel-gray metal is characterized by the highest melting point out of all the discovered elements, highest boiling point, density comparable with uranium or gold, and also low sputtering grade. These properties represent important criteria for which this material is selected as a basic for wall construction in the case of future fusion devices, like International Thermonuclear Experimental Reactor (ITER). In this case, the expected power and temperature of fusion plasma that directly interact with the tungsten components can trigger the wall erosion and formation of dust in the plasma chamber. This dust, composed of tungsten particles of nanometric and micrometric size, can present safety issues in case of failure of confinement during a loss of vacuum accident (LOVA). Accidental interaction of tungsten nanoparticles can be harmful to fusion device workers [12]. In addition, it is known that metallic nano- and microparticles, when dispersed in the ambient air over certain concentrations, are dangerous for health. The concentration of tungsten particles is reported to be small, less than 10 ng/m^3 in normal situation, in ambient air. But, in industrial processes involving the tungsten production, this concentration is alarming and can reach 62.3 mg/m^3. Considering this increased concentration, the risk of inhalation or contamination with these nanoparticles is high. Increased tungsten use could lead to contamination of air, water, and soil near tungsten mines or industrial sites [13]. In speciality studies, exposure of W particles has been shown to have no adverse effect at a concentration of up to 100 μg/mL, and the toxic effect is significant at a concentration of at least 250 μg/mL [1]. For this study, the toxicity was evaluated on liver cells. This cell line is relevant for the toxicity evaluation, because, by ingestion, cells may be exposed to these tungsten particles. However, particles with a diameter of 27 μm were used. Bolt et al. [14] has shown that tungsten increases breast cancer metastasis as a result of its use in medical devices. Thus, studies that present the toxicity of pure tungsten are mostly focused on its elemental presence or microscale particles, while the reports on this material toxicity at nanoscale size are marginal and extremely rare.

1.2 Toxicology studies devoted to tungsten compounds

Considering the hardness of tungsten carbide, this material is of the greatest interest in the industrial field. In order to develop nanoparticles at a large scale in this area, the potential risks on human health and on the environment have begun to be taken into account. The toxicity of tungsten (WC) and tungsten carbide (WC-Co) doped with cobalt nanoparticles was evaluated on a wide range of mammalian cells (lung, skin, and colon cell lines, as well as in neuronal and glial cells). The WC and WC-Co particles in this study have an average particle size of 145 nm. For the toxicity evaluation, the maximum concentration selected for nanoparticle suspensions was 30 μg/mL. WC nanoparticles have no increased toxicity for these cell lines. However, cytotoxicity became major when the particles were doped with Co. The most sensitive to particles were astrocytes and colon epithelial cells. The findings demonstrate that by doping the tungsten carbide nanoparticles with Co significantly increases their cytotoxic effect [15]. Based on the increased toxicity observed, the studies continued with the toxicity tests of Co-doped tungsten carbide nanoparticles, characterized by different sizes and morphologies, on various cell lines.

Thus, the cytotoxicity of the WC-Co particles, which are 40 μm in diameter and both spherical and cortical, was evaluated. The toxicity study was evaluated using two pulmonary cell lines. This study has, in contrast, noted that the toxicity of these particles has no correlation between cytotoxicity with diameter and specific morphology of these nanoparticles [16]. Another study examined the WC-Co particle toxicity with particle size using lung epithelial cells. Thus, nanoparticles (98 nm) and WC-Co microparticles (3.4 μm) have been shown to influence the toxicity in a dose-dependent and exposure time manner. Thus, nano-WC-Co particles have caused significant toxicity compared to micro-WC-Co particles at lower concentrations and shorter exposure times. For the viability test, the cells were exposed to either nano- or microparticles at concentrations up to 1000 μg/mL for exposure periods of time up to 48 hours. It has been observed that WC-Co nanoparticles have been internalized by the pulmonary epithelial cells, suggesting that internalization can play a key role in the increased toxicity of nanoparticles [17].

Furthermore, following all organs at risk in case of inhalation (lungs, liver, and kidneys), the cytotoxicity of WC-Co nanoparticles with a diameter of 60 nm was studied. In this study, the maximum concentration of nanoparticles added in cell medium was 150 μg/mL. These nanoparticles induce cellular mortality, DNA breaks in renal and hepatic cell lines, but do not induce significant cytotoxic effects in lung cells [18]. Based on these studies, WC-Co has been recognized by the National Institute for Occupational Safety and Health as presenting a health hazard to people after inhalation at the workplace. The use of these particles in the industrial field is high, and human health risks remain poorly studied, despite the fact that the International Agency for Research on Cancer (IARC) classifies them as "probably carcinogenic" materials for humans [18].

Beside the tungsten carbide, which was declared toxic in most of these studies, other tungsten alloys were considered relatively inert, and therefore some of these materials did not seem to present a significant risk for human toxicity. However, recent research findings have raised concerns about the possible adverse health effects after acute and chronic exposure to tungsten oxide. Based on unique biophysical properties, this material is considered an important candidate for a broad range of applications, starting with industrial use, with products such as flame-retardant fabrics, X-ray screens, gas sensors, automotive glass, or use as a pigment in ceramics and paints [19–22]. This material also finds application in the biomedical field [23] in the form of biological products like pigments, additives, and analytical agents [24]. Moreover, WO_3 nanoparticles are efficiently used as a contrast agent for computed tomography imaging [24].

Therefore, the toxicity of this material at a nanoscale dimension has begun to be carefully evaluated. It has been reported that these compounds of tungsten were absorbed after oral exposure in both man and rat organisms. Tungsten alloys incorporated into the body have been shown to cause metastatic tumors in rats. Tungsten oxide has been found to accumulate in several organs and/or tissues (kidneys, liver, ovaries, prostate, pancreas, lung, heart, muscles, spleen, and bones) after a single oral dose [25, 26]. Furthermore, at a nanometric scale, recent findings have shown that tungsten oxide nanoparticles have cytotoxic potential [23]. Various in vitro studies have shown that tungsten oxide, at higher concentrations, significantly increases the primary viability hepatocytes of rats [26] (0.3, 0.5 and 1 mg/mL). At the same time, even when a study highlighted nanotoxicity comparison between several materials, acute toxicity was observed in tungsten oxide at concentrations above 0.05–1 mg/mL [4].

Another study regarding tungsten oxide found it has cytotoxic effect on human alveolar epithelial cells, human epithelial colorectal cells, and murine fibroblast cell line at concentrations above 100 μg/mL [11]. Contrary to these, the tungsten oxide nanoparticles are not toxic to these cells at low concentrations (3–100 μg/mL) and to human lung carcinoma (A549) cells [24]. In this article [18], a comparative study between tungsten oxide nanoparticles and microparticles was carried out. It was revealed that WO_3 nanoparticles induced a significant cell viability reduction and, at nanoscale dimension, increased cell membrane damage at higher doses (200 and 300 μg/mL, respectively). These results are in contrast with WO_3 microparticles. WO_3 microparticles did not incite any toxicity attributes in comparison to control samples for the tested concentrations.

1.3 Content of the chapter

After the short review regarding tungsten nanoparticles and the studies that were carried out until now regarding this particular toxicology, it resulted that the knowledge related to the toxicity of tungsten containing particles with tens and hundreds of nanometer size is limited or even missing. This dimension is particularly important because of low efficiency of particle collection by filters at this size scale. Moreover, we did not find any study that reveals the toxicity of tungsten nanoparticles on epithelial cells, despite the prevalence of the absorption of this material through the skin. Taking into account the possible release of tungsten nanoparticles into the atmosphere following a nuclear accident (LOVA) [12], nanoparticles would interact primarily with the epithelial tissue of the workers. This investigation is the further purpose of the present paper; and this is the reason why the normal human skin fibroblast cell line BJ ATCC CRL 2522 was chosen for toxicity evaluation after interaction with tungsten nanoparticles.

In order to undertake the cytotoxicity studies, tungsten dust has been purposely produced using the magnetron sputtering combined with gas aggregation (MS-GAS) technique. Scanning electron microscopy (SEM) and dynamic light scattering (DLS) were used to investigate the nanoparticle morphology and nanoparticle behavior in liquid medium, respectively. This synthesis method and also nanoparticles characterization will be presented in Section 2. In Section 3 toxicology studies will be presented. Doses with different concentrations of tungsten nanoparticles were used. At low concentration of tungsten nanoparticles, the cell viability was investigated using a cell viability assay MTS test ((3-(4, 5-dimethylthiazol-2-yl)-5-(3-carboxymethoxyphenyl)-2-(4-sulfophenyl)-2H-tetrazolium)). In contrast, at a higher concentration of tungsten particles, SEM was used to offer insights into the process of interactions between inorganic nanoparticles and epithelial cells. Also, for every nanoparticle concentration added over the skin cell culture, optical microscopy investigation was accomplished. Finally, in the Conclusions section, we elucidate if tungsten nanoparticles, characterized by a spherical shape and approximately 100 nm in diameter, present high toxicity or not with respect to human skin cells.

2. Nanoparticles synthesis and characterization

2.1 Synthesis of tungsten nanoparticles

Nanoparticles are generally synthesized using different methods largely presented in literature. These are classified mainly in chemical and physical ones [27–29]. In this work we preferred to use a physical synthesis method [30, 31]. Similar with the tokamak device, we have chosen a plasma-based method: magnetron sputtering combined with gas aggregation (MSGA). This method was introduced in the last decade of the last century [32], and starting with that moment, it was used for the synthesis of nanoparticles based on different materials (which includes metals [33] and their compounds [34]).

The schematic figure of the experimental setup is presented in **Figure 1**. The setup assembled from two chambers, an aggregation chamber (MSGA cluster source) which is attached to a vacuum deposition chamber. The cluster source consists in a water-cooled stainless-steel tube in which a magnetron sputtering plasma gun is mounted. The opposite face of the cluster source ends in a conical-shaped flange presenting a 1.5-mm aperture. The distance between the magnetron target (tungsten, 2″ in diameter, purity of 99.95%) and the exit aperture defines the space

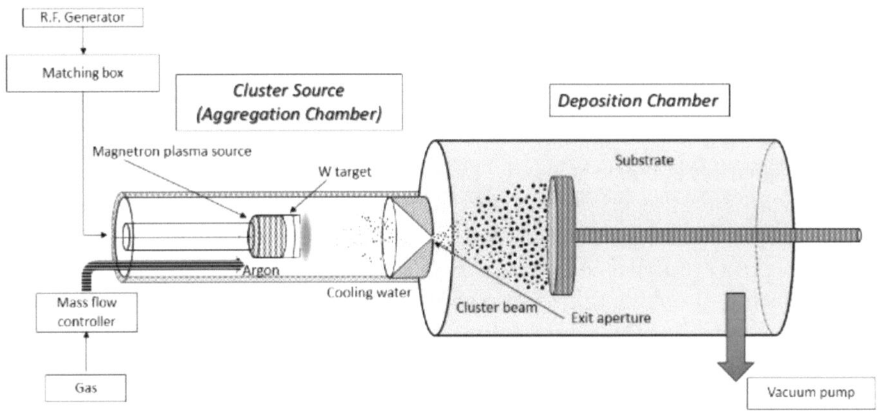

Figure 1.
Schematic of the experimental setup used for synthesis of tungsten nanoparticles.

where the aggregation of nanoparticles takes place; in this work the length of this space was 90 mm.

The following subsystems are integrated into the experimental system: a pumping unit consisting of a turbomolecular pump providing a high vacuum in both chambers and also pressure sensors; and the second subsystem is represented by the gas intake and the mass flow controller (MFC), which helps us control the gas flow in the chambers. In addition, in order to sustain the plasma discharge, a radio-frequency generator, provided with an impedance matching box, was used. The frequency used was 13.56 MHz, and the RF discharge power was constantly 80 W.

The process gas (Ar, 99.9999% purity) is fed directly in the aggregation chamber at a flow of 5 standard cubic centimeter per minutes (sccm), and it is evacuated (via the small aperture) by the pumping system of the deposition chamber. In MSGA system the working gas plays the following roles:

i. To maintain the magnetron plasma discharge and thus, to produce by sputtering the tungsten atoms for nanoparticles growth.

ii. To produce, by three body collisions (W + W + Ar), the initial germs (dimers of W-W type) for further growth of the nanoparticles.

iii. To transport the nanoparticles from the cluster source in the deposition chamber via the exit aperture of the cluster source; this last action is sustained by to the difference of pressure appearing between the deposition chamber (0.5 Pa) and the cluster source chamber (80 Pa).

In accord with the models for nanoparticles growth in magnetron sputtering plasmas presented in the literature [35], the W nanoparticles are formed in the following manner.

Firstly, the three body collisions taking place between two W atoms and an Ar one lead to the formation of W-W dimers. These act like initial germs and increase by further condensation of metallic vapors (process named accretion), until small nanocrystallites are formed. These present dimensions in between 2 and 10 nm. It is important to note that they may be charged; the charging proceeds by the interaction of nanocrystallites with plasma electrons (leading to negative charge) and ions (leading to positive charge). The electrostatic interaction gathers the nanocrystallites leading to formation of nanoparticles. This process is named coagulation.

Finally, the W nanoparticles are transported by the Ar flow in the deposition chamber, which are collected on glass and Si substrates (**Figure 2**). The deposition rate of WNPs was about 5 mg/hour.

2.2 Nanoparticles characterization

Before performing biological tests in order to evaluate the possible toxicity of tungsten nanostructures, aspects of nanoparticle morphology were investigated using SEM after which nanoparticle behavior in liquid medium was evaluated using DLS.

2.2.1 Morphological analysis by SEM

The laboratory model used to evaluate the possible toxic effect of tungsten dust was formed by a powder assembled from nanoparticles. This powder was extracted from the glass substrate and transferred on a carbon tape. The morphology of nanoparticles was investigated by SEM using a Zeiss EVO 50XVP with LaB_6 electron gun operating at 20 kV. **Figure 3** shows that the particles present almost spherical morphology. The diameter of the nanoparticles is about 100 nm.

2.2.2 Particle dispersion analysis by DLS

Nanoparticle behavior in liquid has to be studied as well, because the interaction with cells takes place in liquid phase. For the biological tests, liquid

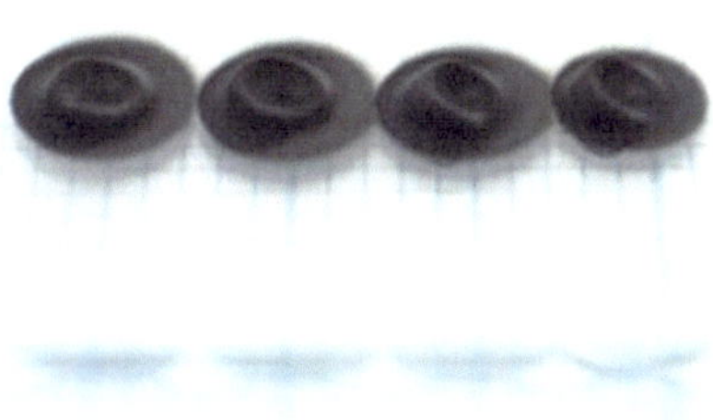

Figure 2.
Tungsten nanoparticle deposits, obtained after four different deposition runs, collected on glass substrate.

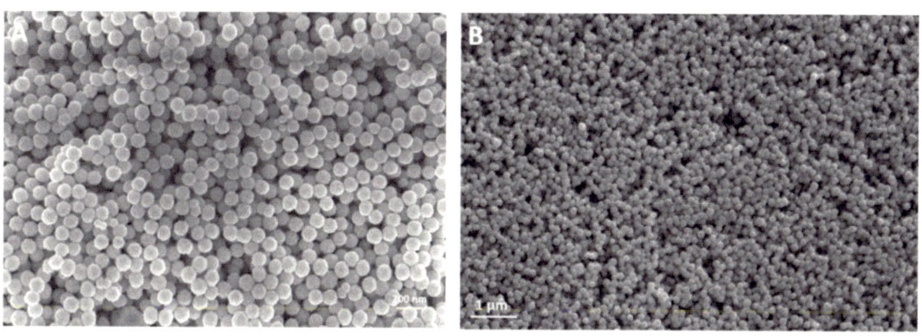

Figure 3.
SEM images of tungsten nanoparticle synthesized by MSGA technique: (A) higher magnification and (B) lower magnification.

dispersions of particles with various concentrations were prepared, as will be detailed in the next section. Such dispersions, consisting of particles dispersed in phosphate-buffered saline (PBS) solutions, were characterized by dynamic light scattering.

Dynamic light scattering (sometimes referred to as photon correlation spectroscopy or quasi-elastic light scattering) is a technique for measuring the size of particles dispersed in liquids, typically in the submicron scale. Dynamic light scattering actually determines the diffusion rate of particles in a particular dispersion medium as a result of the Brownian motion. Brownian motion is the random movement of particles due to the bombardment by the solvent molecules that surround them. Based on this physical process, this light scattering technique allows us to evaluate the apparent size that the particles reach when they are added in a liquid medium. The particles are illuminated with a laser beam with a power of 30 mW, and this light is scattered on the particles. The scattered light is then collected at an angle of 165° for size measurements and then measured with an extremely sensitive detector. After the detection, the hydrodynamic diameter is evaluated, which represents the diameter of a sphere having the same diffusion coefficient as the particle measured, also taking into account the layer of hydrates surrounding the particle/molecule [36].

The data analysis was performed with the DelsaNano 2.21 software. The results, obtained for three dispersion concentrations, used later in the biological tests, are presented in **Figure 4**.

It can be observed that the apparent diameter of nanoparticles in liquid is larger than the particle size indicated by SEM. Also the diameter increases with the concentration of nanoparticles added in liquid, indicating a tendency to agglomeration. The use of an ultrasonication procedure reduces the agglomeration, decreasing the apparent diameter.

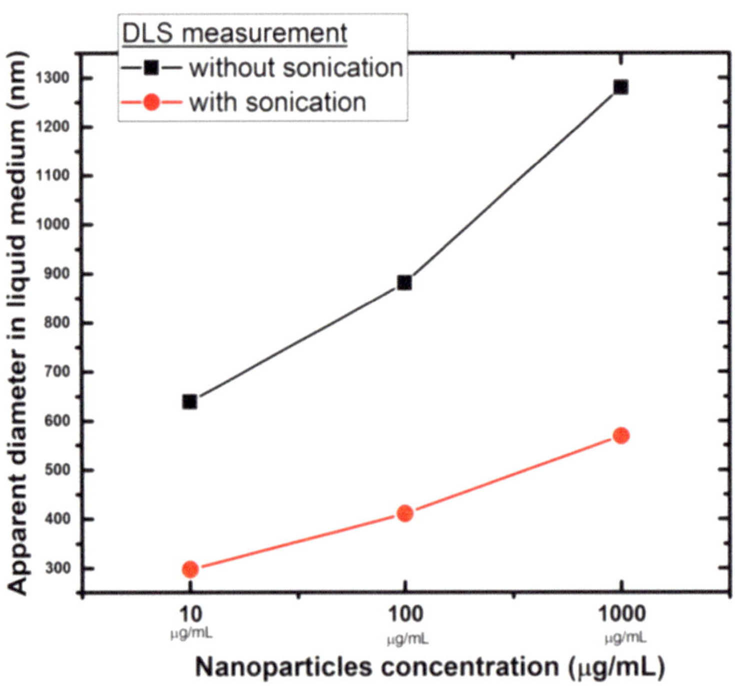

Figure 4.
Apparent diameters resulted from DLS measurements at various dispersion concentrations (the concentrations of 10, 100, and 1000 µg/mL correspond to the samples indexed as C2, C3, and C4 in the Section 3).

3. Evaluation of toxicity

3.1 Cell line selection and preparation of cells samples

For the run of the biological assays, a human skin fibroblast culture, called BJ ATCC, CRL 2522, was used. These cells were derived from an initial ATCC lot (American Type Culture Collection) and were obtained from *Homo sapiens* species. This cell line was kept at 37°C in humidified atmosphere (5% CO_2). Fibroblasts were grown in the Eagle's minimum essential medium (EMEM) environment supplemented with 10% fetal bovine serum (FBS), 100 μg/mL penicillin and 100 μg/mL streptomycin (**Figures 5** and **6**).

3.2 Preparation of the biological samples

A working protocol was set to go from the nanoparticle synthesis (the nanoparticles gathered in circles in **Figure 2**) to the dilutions required to assess the toxicity. In order to achieve this goal, after synthesis, the nanoparticles were first collected from the substrate, weighed, and added in saline solution, forming a stock solution. Before being added over the inoculated cells, the stock solution was sterilized. Furthermore, dilutions were prepared from the stock, which were utilized for toxicity investigations. This protocol is described in **Table 1** and illustrated in **Figure 7**.

In general, a decisive factor in the field of toxicology is determined by daily exposure to nanoparticles and the problem that arises due to it is that in vitro assays cannot be directly correlated with repeated exposure. Therefore, in the present study, in addition to the usual concentrations used for this type of analysis (1, 10, 100 μg/mL), two higher concentrations of nanoparticles were added in EMEM, concentrations that presents a major importance in the case of a nuclear accident, taking into account the purpose of the work [37]. In conclusion, dilutions from the stock suspension were made in order to obtain final concentrations of 1, 10, 100, 1000, and 2000 μg/mL, concentration used to evaluate the possible toxicity effect of tungsten nanoparticles (**Table 2**).

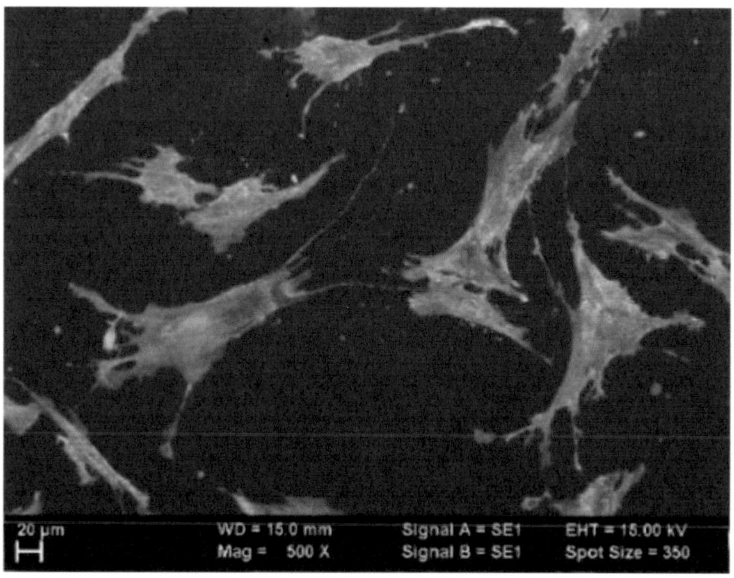

Figure 5.
SEM images of the fibroblast cells after being inoculated for 24 hours.

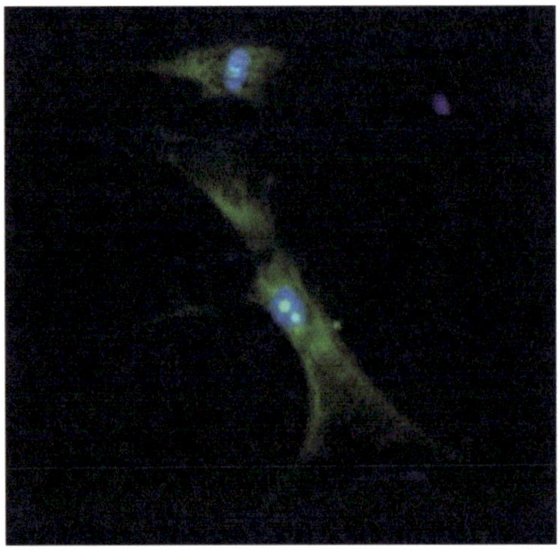

Figure 6.
Cells observed using a fluorescence microscope. Both the nucleus and the cell body are presented.

Preparation of nanoparticle samples to assess cytotoxicity	
Nanoparticles synthesis and collection on substrate	Step 1, **Figure 2**
Sampling nanoparticles from the glass substrates	Step 2, **Figure 7**
Weighing nanoparticle powders	Step 3, **Figure 7**
Preparation of the stock solution (10 mg/mL)	Step 4, **Figure 7**
Solution sterilization	Step 5, **Figure 7**
Performing dilutions, starting from the stock solution	Step 6, **Figure 7**

Table 1.
The steps required to prepare the samples for cytotoxicity evaluation.

At the same time when performing the particle dilutions, the cells were inoculated in the well plate. After the cell attachment on the bottom of the wells, in every well, a different dilution (from C1 up to C5) was added over the cells. This was the first step required to check on the interaction between nanoparticles and skin fibroblast cells, and in the following the meaning of the sample names is control sample for the sample with cells without nanoparticles; blanc for the sample with nanoparticles without cells; and C1, C2, C3, and C4 for the samples containing both cells and nanoparticles.

3.3 Optical microscopy investigation of interaction of cells with nanoparticle dilutions

For a first qualitative analysis of the influence of tungsten particles on dermal fibroblast cells, optical microscopy was used. The evaluation, after each step such as cell seeding in the wells, adding nanoparticle dispersions over the cells inoculated and analyzing beforehand cell viability with the MTS test, respectively, was carried out using the optical microscope inverted Olympus CKX31SF with 10×, 20×, and 40× magnifications.

At low concentrations (C2, C3), the cell attachment to the substrate is still present and the nanoparticle clusters are not well highlighted. We can conclude that cells are still viable in tungsten nanoparticle dispersion.

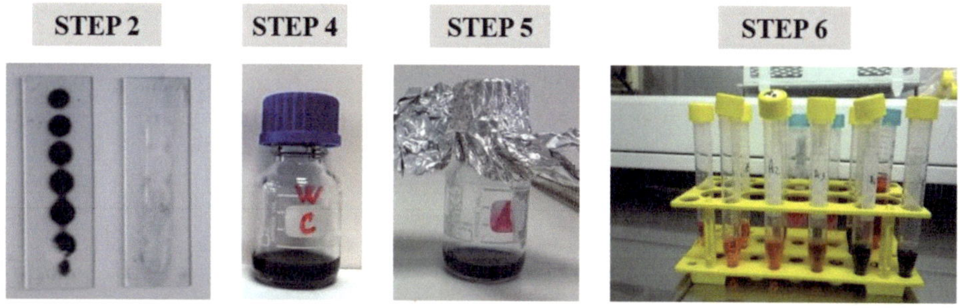

Figure 7.
Images which illustrate the steps that were required for the preparation of nanoparticle samples to assess cytotoxicity.

C1	Low concentration of nanoparticles	1 μg/mL
C2		10 μg/mL
C3		100 μg/mL
C4	Higher concentration of nanoparticles	1000 μg/mL
C5		2000 μg/mL

Table 2.
Dilutions based on stock concentration.

However, for higher concentrations, cells tend to be completely coated with nanoparticles, and moreover, it can be observed how certain cells change their shape, becoming round (a precursor for cell death), but we cannot be sure if the cells are still viable or not.

From the optical microscope images, it is possible to highlight that cells remain anchored on the substrate, even when the concentration of nanoparticles added to the medium is very high. This is an indication for cell viability and a suggestion that, at least in low concentration (<100 μg/mL), tungsten is certainly not a toxic material (**Figure 8**).

3.4 MTS test ((3-(4, 5-dimethylthiazol-2-yl)-5-(3-carboxymethoxyphenyl)-2-(4-sulfophenyl)-2H-tetrazolium))

The MTS ((3-(4,5-dimethylthiazol-2-yl)-5-(3-carboxymethoxyphenyl)-2-(4-sulfophenyl)-2H-tetrazolium)) test is a colorimetric method used for cytotoxicity analysis [38]. This test is used to quantify viable cells. This method is based on the mitochondrial oxidoreductase enzyme reduction by a tetrazolium salt (MTS compound). These enzymes are found only in viable cells—**Figure 9**. Thus, based on this interaction, a formazan compound is generated, and this compound is soluble in the culture environment. The presence of this compound changes the solution absorbance. The presence of the formazan compound is an index of mitochondrial activity, being proportional to the number of viable cells. Formazan exhibits maximum absorbance at 490 nm in a saline solution. Since the MTS reagent is sensitive to light, the test is performed by limiting the ambient light exposure. In addition, at higher concentration, the tungsten nanoparticles also change the dispersion absorbance, and this change interferes with formazan absorption. Therefore, the MTS test does not offer concluding results at high particle concentration, and this is the reason why we performed MTS investigation only for low-concentration samples

(C1, C2, and C3). Another method to evaluate the toxicity for high concentration of nanoparticles is required, and we conclude on this aspect using SEM investigations in Section 3.5.

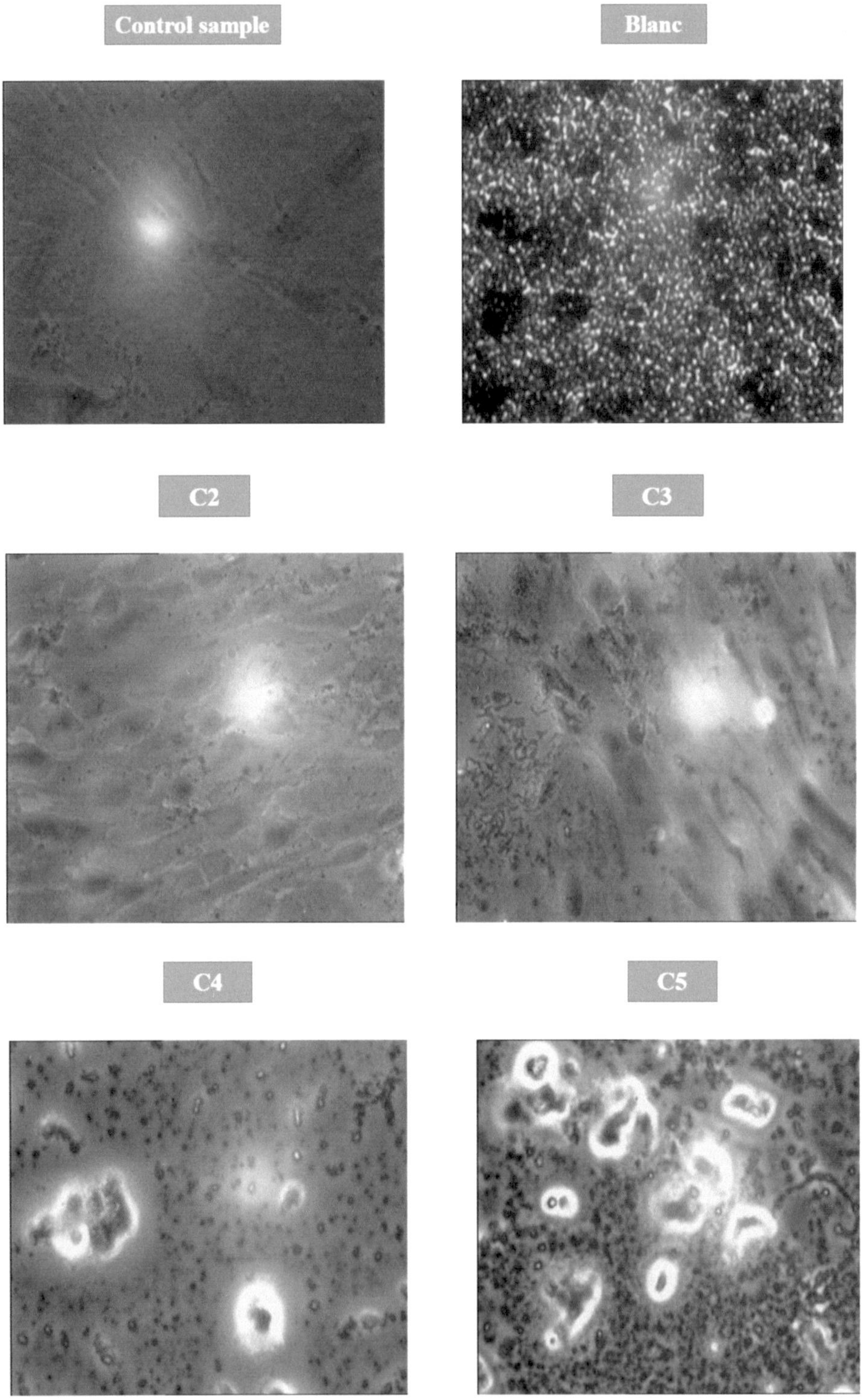

Figure 8.
Optical images of nanoparticle dispersions over the cells inoculated.

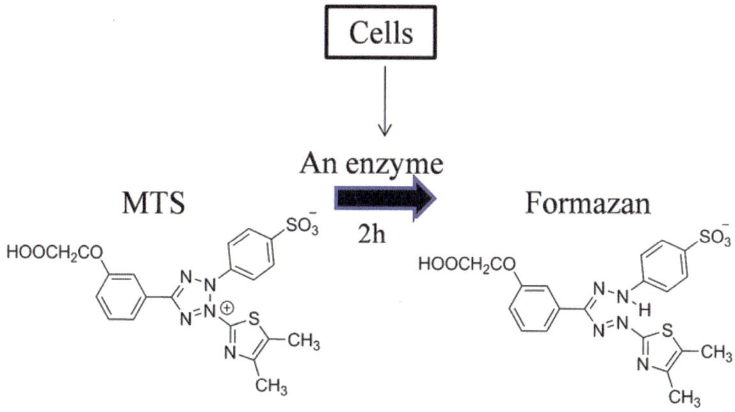

Figure 9.
The chemical structures of the MTS (((3-(4,5-dimethylthiazol-2-yl)-5-(3-carboxymethoxyphenyl)-2-(4-sulfophenyl)-2H-tetrazolium))) reagent and the formazan format compound obtained after the interaction between these compounds with the mitochondrial enzyme.

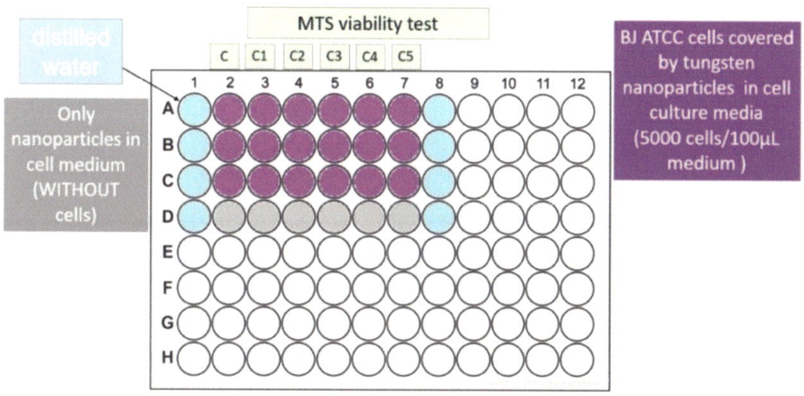

Figure 10.
Wells plate used for the evaluation of MTS assay. In this figure the use of every wells from the plate is highlighted.

Firstly, **Figure 10** shows the order in which different concentrations of tungsten were added over the inoculated cells in the well plates. **Figure 11** shows the MTS effect. MTS was added over dilutions before the cytotoxicity analysis. After 24 and 48 hours, respectively, nanoparticle suspensions were added over the cells inoculated in well plates, and MTS substance was pipetted into each well. This substance was left to act for 3 hours. The 96-well plate, as previously described, was introduced into Mithras spectrophotometer, Berthold Technologies, with the added MTS substance, for each well resulting in an absorbance value.

From the arithmetic average value of the three replicates, corresponding to each concentration (C1–C3), after the absorbance of the blanks was subtracted, the absorbance value was calculated using formula (1). The value of the viability percentage was calculated according to the control sample value, whose viability was considered 100%.

$$\text{Cell viability (C1 up to C3)} = \frac{\text{Absorbance value (C1 up to C3)} * 100}{\text{Control sample absorbance value}} \qquad (1)$$

The MTS results, for 24 and 48 hours of interaction, are presented in **Figure 12**. The cytotoxic effect of WNPs on cells is strongly influenced by the dilution

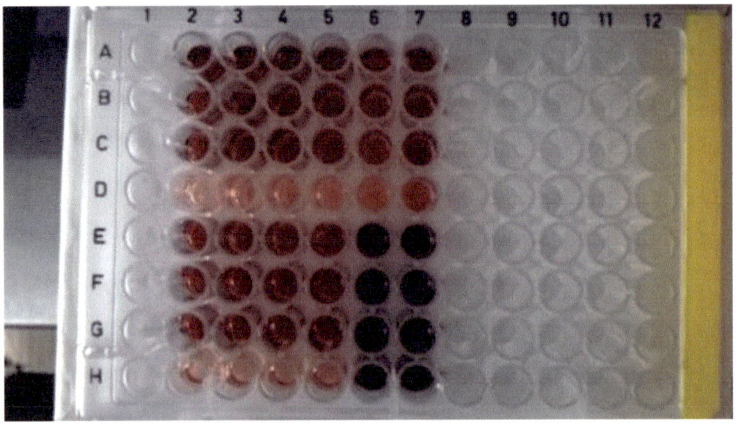

Figure 11.
The 96-well plate in which the MTS solution was put over the cells that were incubated for 24 hours. The formazan presence can be observed in wells by the change of the dispersion color.

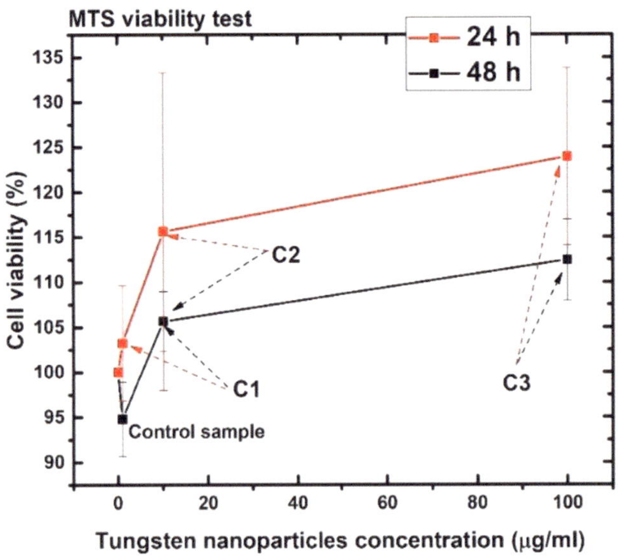

Figure 12.
Results of the MTS cell sample test after 24 and 48 hours, respectively, after the addition of nanoparticles over the anchored cells on the well. It can be seen how the viability decreases after 48 hours.

concentration. According to the results, at low concentrations, the tungsten nanoparticles are not toxic for human skin cells. Nevertheless, cell viability declines after 48 hours in comparison to 24 hours, which indicate that tungsten is not a material that helps cell proliferation.

3.5 Scanning electron microscopy investigations

Scanning electron microscopy was used to obtain complementary information on the process of interaction between the epithelial cells and nanoparticles. In this section, firstly, we will investigate the normal behavior of the cell attachment and cells morphology, in the absence of nanoparticles. In contrary, when different concentrations of nanoparticles were added over the cell culture (from C1 to C5), SEM examination informs us on the changes that can appear in their morphology and also in their viability. These aspects will be discovered using secondary electrons

(SE) imaging. In addition, by operating the instrument in the backscattered electron (BSE) imaging more, details from in-depth of the cell, like the presence of particles beyond the membrane thickness, can be revealed. This process, so-called internalization, is illustrated in **Figure 13**, where images of the same area cell,

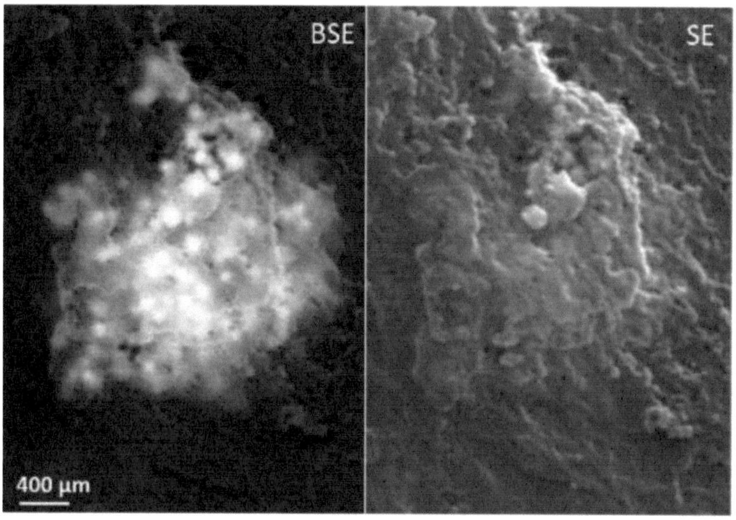

Figure 13.
SE and BSE images are compared and in the BSE image; the internalized particles are highlighted.

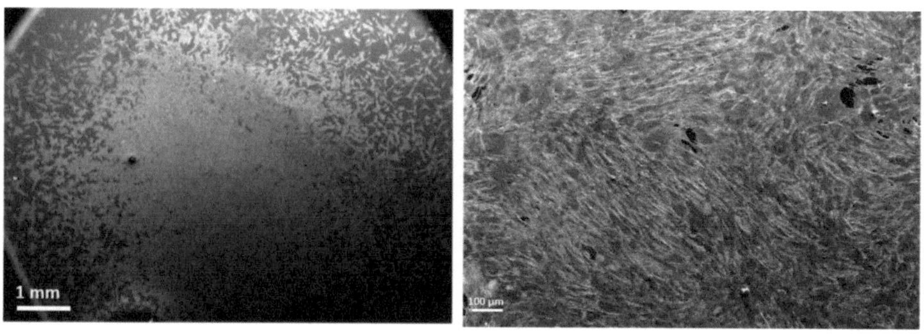

Figure 14.
SEM images, at low and higher magnification, of the control sample, taken using secondary electrons. Normal cellular development can be observed.

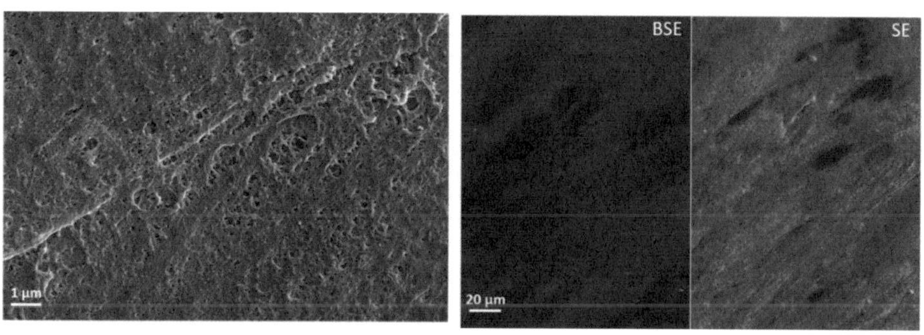

Figure 15.
SEM images taken with the secondary electron (SE) and backscattered electron (BSE) technique of a test sample, where no presence of nanoparticles is observed.

recorded in SE and BSE modes, are compared. Moreover, the backscattered electron technique is based on the contrast between different atomic numbers (Z), so that the nanoparticles with a higher atomic number (92) are brighter in these images than the cell body.

3.5.1 Control sample

In **Figure 14**, normal morphology of the fibroblast cell is observed for the control sample. It can be seen how cells tend to cover the entire surface of the well. An important point highlighted by the higher magnification image is the interconnection between cells. Using the backscattered electron technique (**Figure 15**, BSE compared to SE images), no sign of nanoparticles, or simply just impurity inside the cellular body, or any intrinsic cellular element that could affect the subsequent development of the cells can be observed.

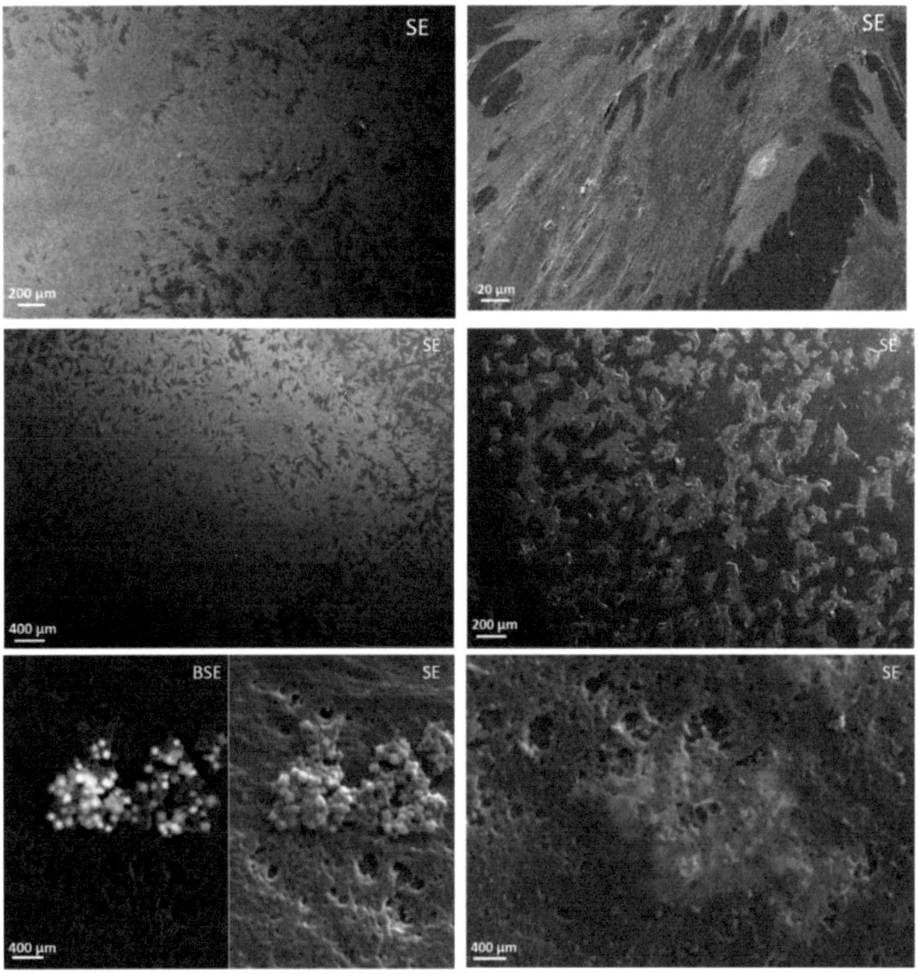

Figure 16.
Images of cells over which a small amount of tungsten nanoparticles (C1—1 µg/mL) were added. Even at such a small concentration added in the cellular environment, the presence of nanoparticles, both on the cell surface and inside the cell body, is observed. At the same time, a difference in the cell number and morphology can be observed in contrast with the control sample.

3.5.2 Low concentration of nanoparticles (samples C1 up to C3)

When a small amount of nanoparticles is added over the pre-inoculated cells, it can be observed first that the bonding between them is not destroyed and secondly, the nanoparticle presence does not radically modify cell morphology. However, some of them become slightly circular, and the number of attached cells on the well slightly decreases. Thus, at a low concentration, tungsten nanoparticles do not dramatically affect cellular proliferation, implicitly cellular viability, as it is demonstrated also by the MTS test. Nevertheless, a nanoparticle effect was observed. C1 nanoparticle concentration effects are present in **Figure 16**, while in **Figure 17** the C3 effects are highlighted.

3.5.3 Higher concentration of nanoparticles (C4 and C5)

At higher concentrations of nanoparticles added over the cells that were previously attached to the substrate, one can first notice a radical change in the cell morphology. They tend to become round, a sign that precedes cell death. At the same time, the number of cells attached on the walls decreased in contrast with the control sample, which shows that the nanoparticles obviously affect the proliferation of fibroblast cells, effects presented in **Figures 18** and **19** for

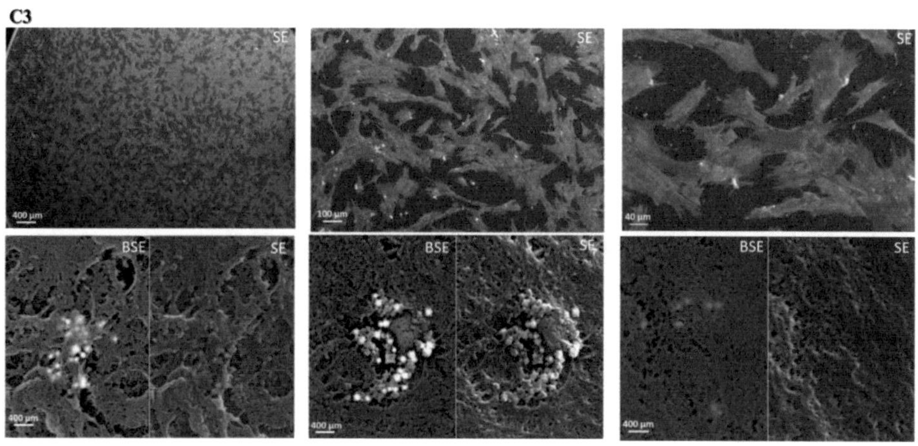

Figure 17.
When the amount of nanoparticles added over attached cells increases (C3—100 μg/mL), a decrease in the cell number on the well surface can be seen, and the number of nanoparticles on the cell surface, as well as the number of nanoparticles internalized in the body of the cells, is higher.

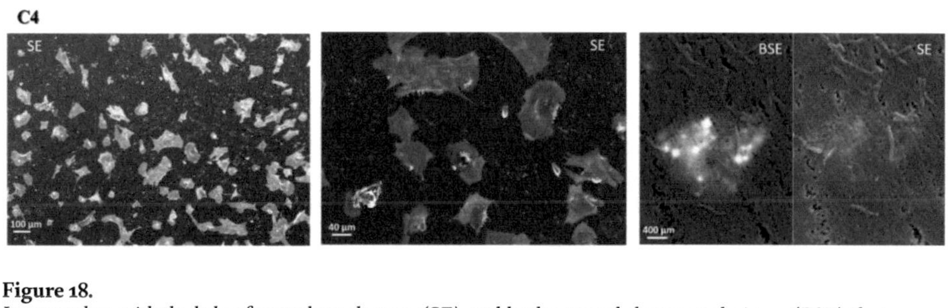

Figure 18.
Images taken with the help of secondary electron (SE) and backscattered electron techniques (BSE) that highlight the changes in cell morphology after interaction with tungsten nanoparticles.

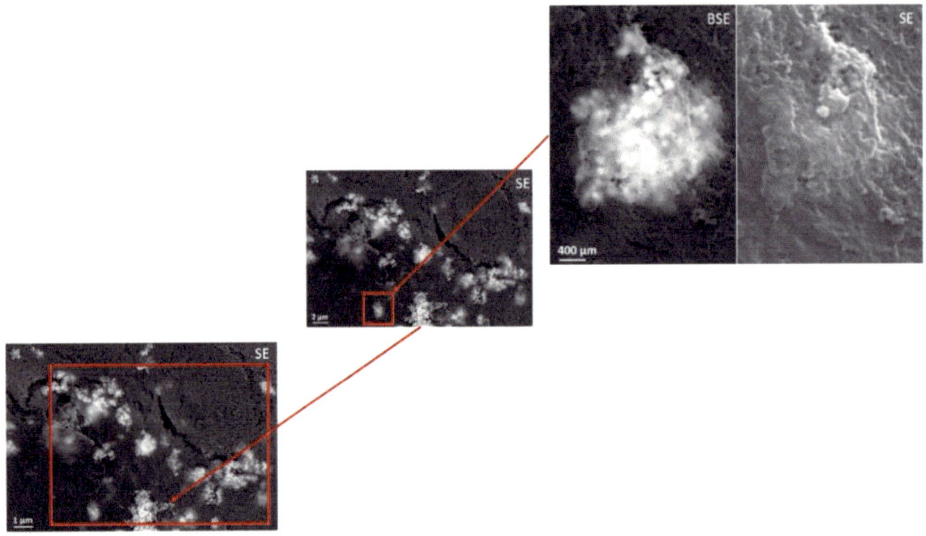

Figure 19.
Images taken with the help of secondary electron (SE) and backscattered electron techniques (BSE) at various magnifications that capture the effect of nanoparticles when large amounts of them were added over inoculated fibroblast cells.

C5

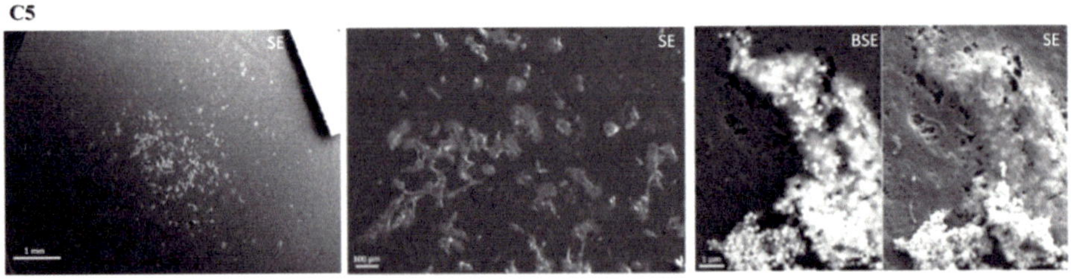

Figure 20.
Electron microscopy images at various magnifications that capture the effect of nanoparticles when a very high concentration of cellular media mixed with nanoparticles is added over the inoculated fibroblast cells. These nanoparticles influence the growth of cells because they remain on the surface, but at the same time, some of them also penetrate the cellular body, beneath the membrane.

C4 concentration and **Figures 20** and **21**, for C5 concentration, respectively. At a higher magnification, it is noticeable that the nanoparticles were internalized under the cell membrane, an effect that is highlighted using the backscattered electron technique.

This section shows that no major difference can be seen between the control sample, in which cells have grown normally and the sample of cells over which just a small amount of nanoparticles (e.g., sample C1) has been added. In contrast, when nanoparticle concentration increased, the number of viable cells attached on the well decreased, this being especially noticeable at a low magnification. However, when the magnification increases, it can be observed that nanoparticles cover approximately the entire cell surface, and moreover, they penetrate the membrane, being internalized in the cells (**Figure 21**). This could be observed with back-scattered electrons that nanoparticles can penetrate the cells at a depth of about 100 nm, meaning within the cytoplasm (under the membrane, this being defined by the lipid bilayer having the thickness about 10 nm).

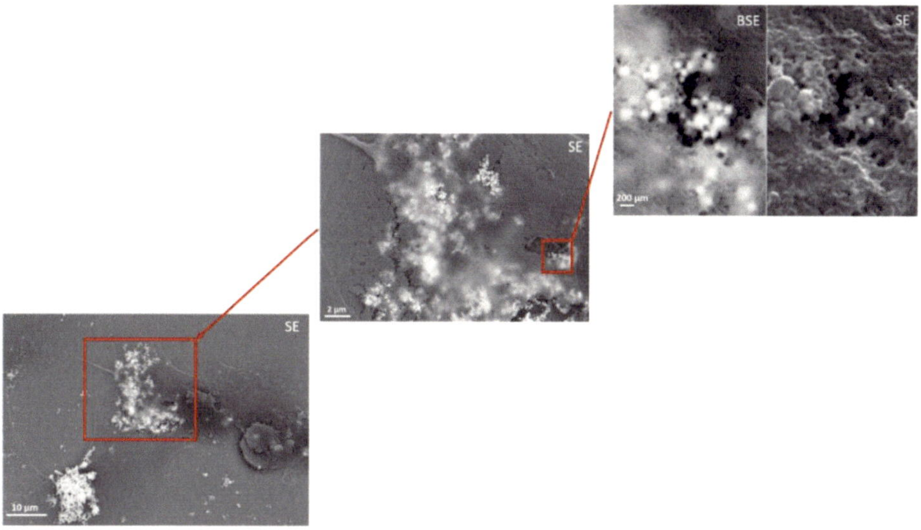

Figure 21.
Images taken using scanning electron microscopy at various magnifications, observing C5 concentration of nanoparticles (2 mg/mL), both on the surface of the cell body and nanoparticles that are internalized under the membrane. These latter nanoparticles can be observed through the backscattered electron technique.

4. Conclusions

This chapter addresses a high-profile scientific field in which we evaluate the possible cytotoxic effects of tungsten nanoparticles through their in vitro exposure. Firstly, according to the chosen research theme, a laboratory model was used in order to investigate the possible toxicity of the tungsten dust. This laboratory model was used to simulate the nanoparticles resulting from a nuclear fusion reactor operation. This model is based on nanoparticle synthesis using the magnetron sputtering with the gas aggregation technique.

This chapter specifically focuses on assessing the toxicity of tungsten particles on fibroblast cells. The toxicity of tungsten powder in dispersion was tested using dermal fibroblast cells. For this investigation, different concentrations of nanoparticles added to the previously inoculated cells in the wells were used. Thus, at low concentrations of nanoparticles added in suspension, this material does not exhibit a high level of toxicity. However, when the nanoparticle concentration added to the cellular medium increases above 100 µg/mL, reaching even 1 and 2 mg/mL doses of interest in accordance with to the concentration of nanoparticles released following a possible nuclear accident LOVA), the toxicity of tungsten is high. This toxicity is primarily observed by significantly reducing the number of viable cells remaining on the substrate. Thus, the ability to multiply in monolayer is lost due to the nanoparticle destruction of the contact between the neighboring cells. Furthermore, loss of cell contact with the surface of the well can be observed. Thus, individual cells appear on the surface. At the same time, optical investigation has observed the change in cell morphology. They tend to have a spherical shape, by retracting their pseudopods. These changes occurring in dermal fibroblast cells after interaction with tungsten nanoparticles are characteristic of the post-apoptotic stage.

Acknowledgements

Part of this work has been carried out within the framework of the Eurofusion consortium and has received funding from the Euratom research and training programme 2014–2018 and 2019–2020 (work package WPEDU-RO, ctr. 1EU-12) under grant agreement No 633053. The views and opinions expressed herein do not necessarily reflect those of the European Commission. In addition, we acknowledge the support in the frame of IFA-CEA project C5-07/2016 and Core Programme 2019 at INFLPR.

Conflict of interest

The authors declare that there is no conflict of interest.

Notes/thanks/other declarations

Author details

Lavinia Gabriela Carpen[1,2], Tomy Acsente[1], Maria Adriana Acasandrei[3], Elena Matei[4], Claudia Gabriela Chilom[2], Diana Iulia Savu[3] and Gheorghe Dinescu[1,2]*

1 National Institute for Laser, Plasma and Radiation Physics (INFLPR), Bucharest, Romania

2 Faculty of Physics, University of Bucharest, Bucharest, Romania

3 Horia Hulubei National Institute for R & D in Physics and Nuclear Engineering (IFIN-HH), Bucharest, Romania

4 National Institute of Materials Physics (NIMP), Bucharest, Romania

*Address all correspondence to: dinescug@infim.ro

References

[1] Hussain SM, Hess KL, Gearhart JM, Geiss KT, Schlager JJ. In vitro toxicity of nanoparticles in BRL 3A rat liver cells. Toxicology in Vitro. 2005;**19**:975-983. DOI: 10.1016/j.tiv.2005.06.034

[2] Khanna P, Ong C, Bay BH, Baeg GH. Nanotoxicity: An interplay of oxidative stress, inflammation and cell death. Nanomaterials. 2015;**5**:1163-1180. DOI: 10.3390/nano5031163

[3] Zhao Y, Wang B, Feng W, Bai C. Nanotoxicology: Toxicological and biological activities of nanomaterials. In: Encyclopedia of Life Support Systems (EOLSS). 2011. p. 26

[4] Jain K, Kohli E, Prasad D, Kamal SSK, Haussain SM, Singh SB. In vitro cytotoxicity assessment of metal oxide nanoparticle. Nanomedicine and Nanobiology. 2014;**1**(1):10-19. DOI: 10.1166/nmb.2014.1003

[5] Grabinski C, Hussain S, Lafdi K, Braydich-Stolle L, Schlager J. Effect of particle dimension on biocompatibility of carbon nanomaterials. Carbon. 2007;**45**(14):2828-2835. DOI: 10.1016/j.carbon.2007.08.039

[6] Li X, Liu W, Sun L, Aifantis KE, Yu B, Fan Y, et al. Effects of physicochemical properties of nanomaterials on their toxicity. Journal of Biomedical Materials Research. 2015;**103**(7):2499-2507. DOI: 10.1002/jbm.a.35384

[7] Proksch E, Brandner JM, Jensen JM. The skin: An indispensable barrier. Experimental Dermatology. 2008;**17**(12):1063-1072. DOI: 10.1111/j.1600-0625.2008.00786.x

[8] Bennat C, Müller-Goymann CC. Skin penetration and stabilization of formulations containing microfine titanium dioxide as physical UV filter. International Journal of Cosmetic Science. 2000;**22**(4):271-283. DOI: 10.1046/j.1467-2494.2000.00009.x

[9] Trop M, Novak M, Rodl S, Hellbom B, Kroell W, Goessler W. Silver-coated dressing acticoat caused raised liver enzymes and argyria-like symptoms in burn patient. The Journal of Trauma Injury Infection and Critical Care. 2006;**60**(3):648-658. DOI: 10.1097/01.ta.0000208126.22089.b6

[10] Tinkle SS, Antonini JM, Rich BA, Roberts JR, Salmen R, DePree K, et al. Skin as a route of exposure and sensitization in chronic beryllium disease. Environmental Health Perspectives. 2003;**111**:1202-1208. DOI: 10.1289/ehp.5999

[11] Ivask A, Titma T, Visnapuu M, Vija H, Käkinen A, Sihtmäe M, et al. Toxicity of 11 metal oxide nanoparticles to three mammalian cell types in vitro. Current Topics in Medicinal Chemistry. 2015;**15**(18):1914-1929. DOI: 10.2174/1568026615666150506150109

[12] Malizia A, Poggi LA, Ciparisse J-F, Rossi R, Bellecci C, Gaudi P. A: Review of dangerous dust in fusion reactors: From its creation to its resuspension in case of LOCA and LOVA. Energies. 2016;**9**:578. DOI: 10.3390/en9080578

[13] Prajapati MV, Adebolu OO, Morrow BM, Cerreta JM. Evaluation of pulmonary response to inhaled tungsten (IV) oxide nanoparticles in golden Syrian hamster. Experimental Biology and Medicine. 2017;**242**:29-44. DOI: 10.1177/1535370216665173

[14] Bolt AM, Sabourin V, Molina MF, Police AM, Negro Silva LF, Plourde D, et al. Tungsten targets the tumor microenvironment to enhance breast cancer metastasis. Toxicological Sciences. 2015;**143**(1):165-177. DOI: 10.1093/toxsci/kfu219

[15] Bastian S, Busch W, Kühnel D, Springer A, Meißner T, Holke R, et al. Toxicity of tungsten carbide and cobalt doped tungsten carbide nanoparticles in mammalian cells in vitro. Environmental Health Perspectives. 2009;**117**(4):530-536. DOI: 10.1289/ehp.0800121

[16] Lanone S, Rogerieux F, Geys J, Dupont A, Maillot-Marechal E, Boczkowski J, et al. Comparative toxicity of 24 manufactured nanoparticles in human alveolar epithelial and macrophage cell lines. Particle and Fibre Toxicology. 2009;**6**:12. DOI: 10.1186/1743-8977-6-14

[17] Armstead AL, Arena CB, Li B. Exploring the potential role of tungsten carbide cobalt (WC-Co) nanoparticle internalization in observed toxicity toward lung epithelial cells in vitro. Toxicology and Applied Pharmacology. 2014;**278**(1):1-8. DOI: 10.1016/j.taap.2014.04.008

[18] Paget V, Moche H, Kortulewski T, Grall R, Irbah L, Nesslany F, et al. Human cell line-dependent WC-Co nanoparticle cytotoxicity and genotoxicity: A key role of ROS production. Toxicological Sciences. 2015;**143**(2):385-397. DOI: 10.1093/toxsci/kfu238

[19] Lassner E, Schubert W-D. Tungsten-Properties, Chemistry, Technology of the Element, Alloys, and Chemical Compounds. 1st ed. New York, US: Springer; 1999. p. 422. DOI: 10.1007/978-1-4615-4907-9

[20] Lee WJ, Fang YK, Ho J-J, Hsieh W-T, Ting S-F, Huang D, et al. Effects of surface porosity on tungsten trioxide (WO3) films electrochromic performance. Journal of Electronic Materials. 2000;**29**(2):183-187. DOI: 10.1007/s11664-000-0139-8

[21] Williams DE, Aliwell SR, Pratt KFE, Caruana DJ, Jones RL, Cox RA, et al. Modelling the response of a tungsten oxide semiconductor as a gas sensor for the measurement of ozone. Measurement Science and Technology. 2002;**13**(6):923-931. DOI: 10.1088/0957-0233/13/6/314

[22] Patnaik P. Handbook of Inorganic Chemicals. 1st ed. McGraw-Hill Professional: USA; 2002. p. 1086

[23] Akbaba BG, Turkez H, Sonmez E, Akbaba U, Aydın E, Tatar A, et al. In vitro genotoxicity evaluation of tungsten (VI) oxide nanopowder using human lymphocytes. Biomedical Research. 2016;**27**(1):229-234

[24] Chinde S, Poornachandra Y, Panyala A, Kumari SI, Yerramsetty S, Adicherla H, et al. Comparative study of cyto- and genotoxic potential with mechanistic insights of tungsten oxide nano- and microparticles in lung carcinoma cells. Journal of Applied Toxicology. 2018;**38**:896-913. DOI: 10.1002/jat.3598

[25] McInturf SM, Bekkedal MY, Wilfong E, Arfstend D, Gunasekar PG, Chapman GD. Neurobehavioral effects of sodium tungstate exposure on rats and their progeny. Neurotoxicology and Teratology. 2008;**30**(6):455-461. DOI: 10.1016/j.ntt.2008.07.003

[26] Turkez H, Sonmez E, Turkez O, Mokhtar YI, Di Stefano A, Turgut G. The risk evaluation of tungsten oxide nanoparticles in cultured rat liver cells for its safe applications in nanotechnology. Brazilian Archives of Biology and Technology. 2014;**57**(4):532-541. DOI: 10.1590/S1516-89132014005000021

[27] Lazea Stoyanova A, Vlad A, Vlaicu AM, Teodorescu VS, Dinescu G. Synthesis of copper particles by non-thermal atmospheric pressure plasma jet. Plasma Processes and Polymers. 2015;**12**(8). DOI: 10.1002/ppap.201400197

[28] Lazea-Stoyanova A, Enculescu M, Vizireanu S, Marascu V, Dinescu G. Effects of process parameters on growth of metal particles by atmospheric pressure plasma jet. Digest Journal of Nanomaterials and Biostructures. 2014;**9**(3):1241-1247

[29] Marascu V, Lazea-Stoyanova A, Stancu C, Dinescu G. The influence of plasma operation parameters on synthesis process of copper particles at atmospheric pressure. Plasma Processes and Polymers. 2018;**15**(1). DOI: 10.1002/ppap.201700091

[30] Acsente T, Negrea RF, Nistor LC, Logofatu C, Matei E, Birjega R, et al. Synthesis of flower-like tungsten nanoparticles by magnetron sputtering combined with gas aggregation. The European Physical Journal. 2015;**69**(6):7. DOI: 10.1140/epjd/e2015-60097-4

[31] Acsente T, Negrea RF, Nistor LC, Matei E, Grisolia C, Birjega R, et al. Tungsten nanoparticles with controlled shape and crystallinity obtained by magnetron sputtering and gas aggregation. Materials Letters. 2015;**200**:121-124. DOI: 10.1016/j.matlet.2017.04.105

[32] Haberland H, Karrais M, Mall M, Thurner Y. Thin films from energetic cluster impact: A feasibility study. Journal of Vacuum Science and Technology. 1992;**A10**:3266. DOI: 10.1116/1.577853

[33] Haberland H, Karrais M, Mall M. A new type of cluster and cluster ion source. Zeitschrift fur Physik D: Atoms, Molecules and Clusters. 1991;**20**(1):413-415. DOI: 10.1007/BF01544025

[34] Shelemin A, Kylián O, Hanuš J, Choukourov A, Melnichuk I, Serov A, et al. Preparation of metal oxide nanoparticles by gas aggregation cluster source. Vacuum. 2015;**A120**:162-169. DOI: 10.1016/j.vacuum.2015.07.008

[35] Kashtanov PV, Smirnov BM, Hippler R. Magnetron plasma and nanotechnology. Physics-Uspekhi. 2007;**50**(5):455-488. DOI: 10.1070/PU2007v050n05ABEH006138

[36] Tantra R, Schulze P, Quincey P. Effect of nanoparticle concentration on zeta-potential measurement results and reproducibility. Particuology. 2010;**8**:279-285. DOI: 10.1016/j.partic.2010.01.003

[37] Drasler B, Sayre P, Steinhäuser KG, Petri-Fink A, Rothen-Rutishauser B. In vitro approaches to assess the hazard of nanomaterials. NanoImpact. 2017;**8**:99-116. DOI: 10.1016/j.impact.2017.08.002

[38] Cory AH, Owen TC, Barltrop JA, Cory JG. Use of an aqueous soluble tetrazolium/formazan assay for cell growth assays in culture. Cancer Communications. 1991;**3**(7):207-212. DOI: 10.1016/0022-1759(93)90092-L

Section 4

Modulation of the Nanomaterial's Toxicity

Chapter 6

Biofunctionalized Polymer Nanomaterials: Implications on Shapes and Sizes

F.F. Razura-Carmona, G.A. Prado-Guzmán, A. Perez-Larios, M.V. Ramírez-Marez, M. Herrera-Martínez and Jorge Alberto Sánchez-Burgos

Abstract

Nanotechnology has been one of the most widely used tools in various industries such as pharmaceutical, food, and chemistry, among others, for the encapsulation of compounds or even microorganisms. However, an analysis of the methodologies or polymer matrices to be used is rarely generated, and these in turn contribute to the objective of the product that is intended to be designed. In addition to the evaluation of its physicochemical, optical, and rheological characteristics, and others, are a set of technological tools that allow predicting the stability of a colloid, however, some of the factors that have less importance as the effect of the synthesis process on the shape and size that a structure may have, studies have been carried out to evaluate this phenomenon, which has become a determining factor in the design of any nanoscale material.

Keywords: functionalization, synthesis, polymers, biomaterials

1. Introduction

Recently, the science of nanomaterials has gained strength, due to the wide applications as it is in the pharmaceutical, food, agricultural and other industries, since it has been a tool that allows making the objectives of each of them more efficient in comparison with the direct compound applications, because there are different processes to trap them generating this form particles resistant to biochemical oxidations, UV radiation, gastric digestion or even in some cases can be taken as transport [1, 2].

In this context, the advances in technologies for encapsulation has allowed the development and implementation of polymer matrices with enhanced properties, however, in some cases the size and amount of the materials are not adequate causing allow stability in the emulsion [3].

Due to the growing demand for these materials and the large amount of study that exists, the custom synthesis has been generated, which allows us to obtain materials in suitable sizes depending on the characteristics of the objective, through basic techniques and in some cases with the basic elements that are described below.

2. Functionalization of polymers: synthesis-controlled shapes and sizes

One of the principal objectives of nanoscience is building small structures for the design of advanced material nanodispositives of high performance. Inorganic nanoparticles are particularly attractive as construction parts for such purposes due to their unique optical, electronegative, and catalytic properties, many of those can be modulated simply by changing their shape, state, or functionality of the surface of the nanoparticle, without changing the composition of the material [4–6].

Figure 1 shows a proposed diagram for some of the sections that encompass nanoscience.

Due to their physical and chemical properties, nanoparticles are often described as artificial atoms. The advances in the processes of synthesis have allowed the precise control over the structural parameters that govern the formation of the nanoparticles which has allowed to adapt the properties of these artificial atoms according to their specific use. The synthesis and modular assembly of nanoparticles allows to exploit their unique properties, which can lead to new applications in catalysis, electronics, photonics, magnetism, as well as chemical and biological sensing.

The nanoparticles can be obtained mainly by two different ways: "Top-down" (Chemical vapor deposition, Thermal evaporation, and Ion implantation) and "Bottom-up" (Microwave irradiation, Colloidal, Solvothermal, Dendrimers, and Sol-gel), where each method exhibit its particular characteristics as described below [7].

2.1 Top-down methods

Top-down focuses on the division of mass solids in small proportions, in which operations such as grinding, chemical methods, or volatilization of a solid can be involved from individual molecules [8]. The most representative is shown below:

- **Chemical vapor deposition** (**CVD**): It consists of the decomposition of one or more volatile compounds, inside a vacuum chamber, to produce products of high purity and yield in solid materials. It is used for the design of mono and polycrystalline materials, amorphous and epitaxial in titanium nitride, carbon nanotubes, silica dioxide, carbon fibers, and others [6].

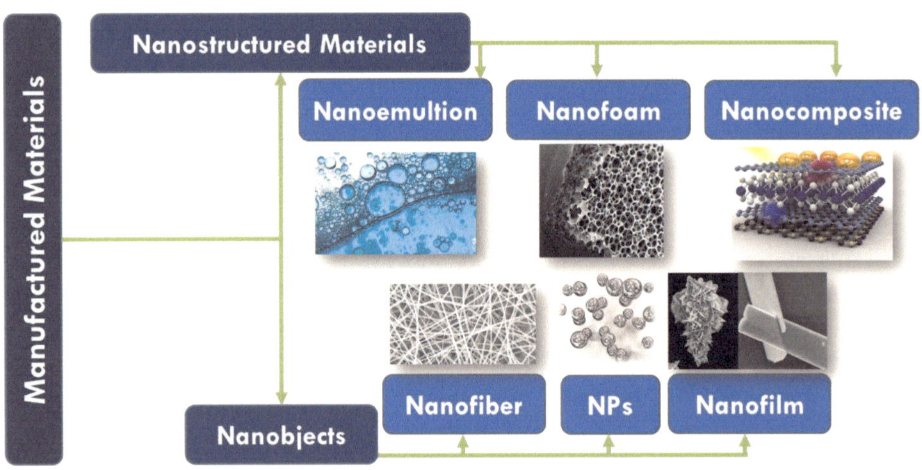

Figure 1.
Diagram of manufactured materials.

- **Thermal evaporation**: It consists of heating until the evaporation of the material that is to be deposited. It is carried out in a vacuum chamber in which the vapor is condensed on a cold sheet, requiring always an accurate control of the growth conditions so as not to produce a modification of the morphology of the deposited layer [9].

- **Ion implantation**: The process consists in the interaction of the ion of a solid material when implanted in another, changing therefore its physical properties. It is used in the manufacture of coating devices for some metals used as semiconductors, as in research in the science of nanomaterials, due to its versatility and ease of control, which allows the synthesis of a wide range of nanocrystals. The ions cause chemical and structural changes in the matrix of origin, since they can be of an element different from the one that composes it, which can be damaged by the glass or even destroyed [10].

2.2 Bottom-up methods

Bottom-up synthesis requires complex, and expensive instrumentation, the most used are under chemical procedures; which is based in the reduction of metallic ions to metal atoms, consequently the aggregation of the atoms in controlled. On the other hand, the nanoparticles obtained by this method exhibit uniform and small structures as described below.

- **Microwave irradiation**: Microwave irradiation produces nanoparticles, which have a low size dispersion without precise control in morphology, as it happens in most "bottom-up" techniques. This method creates a high-frequency electric field, any material can be heated, containing electrical charges such as polar molecules in a solvent or conductive ion in a solid. The polar solvents are heated, and their molecular components are forced to rotate with the field and lose energy in collisions. The conductive and semiconductor samples are heated when the ions and electrons contained in them form an electric current, and the energy is lost due to the electrical resistance of the component. In recent years, given that it is a uniform, effective, and fast method, it has been used as an attractive alternative for the synthesis of nanometer-scale materials. Colloidal nanoparticles of Pt, Ru, Ag, and Pd stabilized by polymers have been prepared by microwave heating, from the metal precursor salts dissolved in ethylene glycol solutions [11] (**Figure 2**).

- **Colloidal**: Colloids are individual particles; they are larger than atomic sizes, but small enough to produce Brownian motion. When the particles are sufficiently large, their dynamic behavior in suspension as a function of time is governed by the forces of gravity and will cause the phenomenon of sedimentation; it is attributed to collective bombardments of a multitude of thermally agitated molecules in a liquid suspension when they are small enough to be colloids. This method is efficient due to the size range that the nanoparticles are produced, which oscillates below 300 nm depending on parameters of elaboration [12, 13] (**Figure 3**).

- **Solvothermal**: It is an efficient method of production of materials consisting of the conjugation of substances that are reacted in a hermetically sealed container at low temperatures (generally at 200°C) and controlled pressure. The objective of the technique is to mix immiscible compounds with each other under normal conditions (aluminosilicates, titans, sulfides, etc.). The

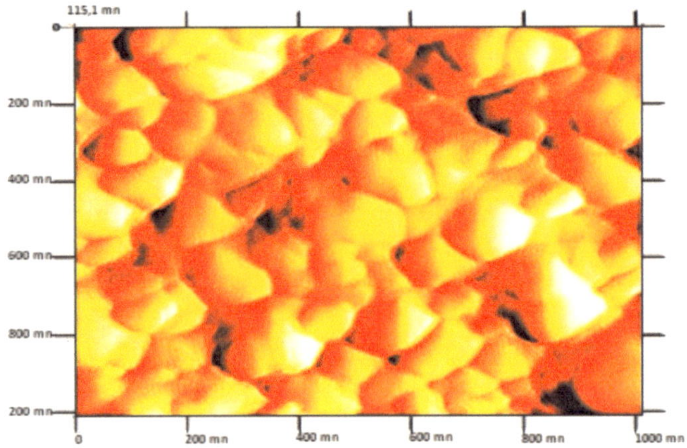

Figure 2.
Atomic force microscopy image of CdS synthesized by microwave heating [9].

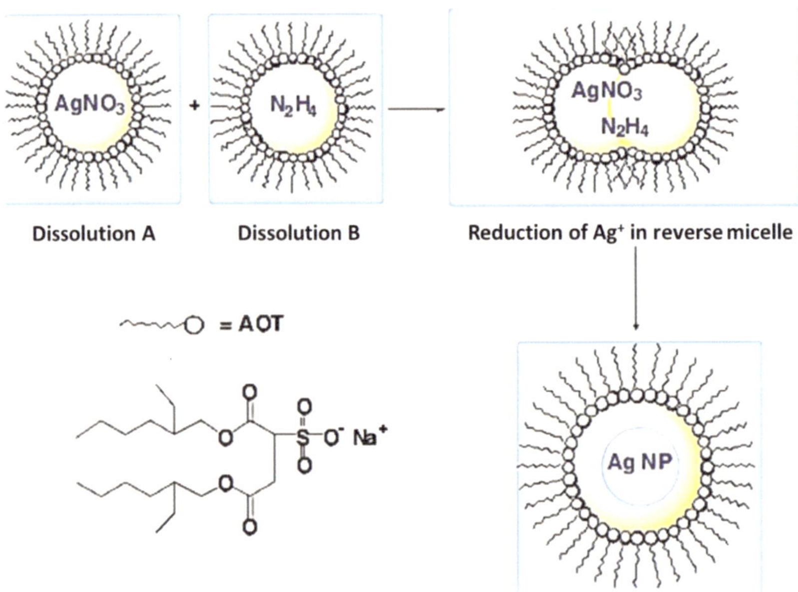

Figure 3.
Synthesis of silver nanoparticles using reversed micelles of AOT (sodium bis (2-ethylhexyl) sulfosuccinate) formed in a dodecane/water mixture [13].

environmental pressure inside the system is exceeded since it occurs in a closed system, in which solvents surpass boiling points, generating a state of "critical" fluid in which certain chemical reactions that not occur under usual conditions. Once the system is cold, the structures reach the crystalline phase. Low energy, accelerated interactions between species, controlled stoichiometry, the greater power of dissolution and transport of reagents, better control of morphology, and a reaction yield close to 100% are some of the advantages of this technique. It is currently used to synthesize a variety of compounds of scientific and industrial interest [14].

- **Dendrimers**: It is a method based on the intervention of micelles, emulsions, and dendrimers, which allows defined shapes and sizes: however, it is a technique that requires a long process to obtain good results in addition to intervening variables

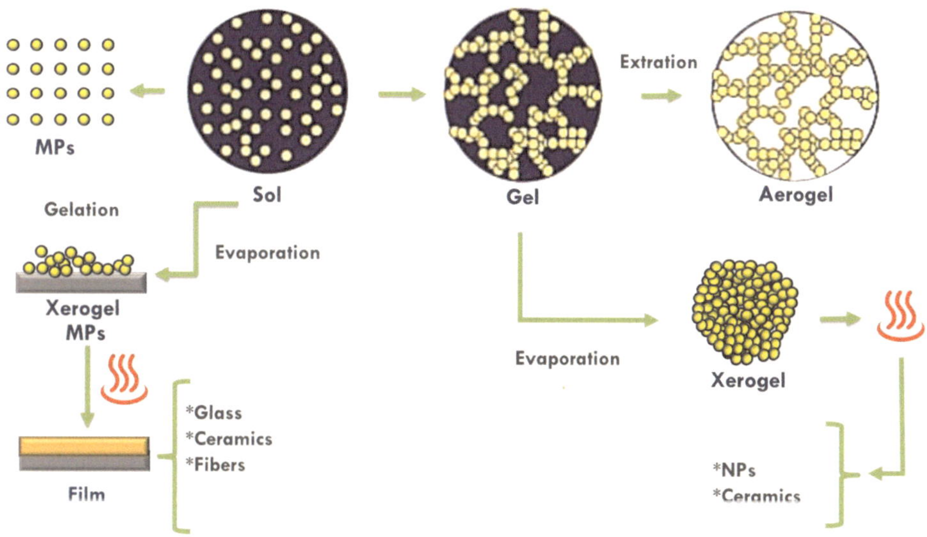

Figure 4.
Stages for obtaining microparticles (MPs) and nanoparticles (NPs) by sol-gel method.

such as time, temperature, pressure, speed of homogenization, and molecular weight, among others, to obtain good results [15].

The most suitable dendrimers for the synthesis of hybrid and monometallic nanoparticles are of different generations with some functional groups. The poly (amidoamines) are the most popular dendrimers used to synthesize gold nanoparticles oscillating between 1 and 3 nm; studies with palladium and platinum with nanomatic characteristics inferior to gold have also been highlighted [16, 17].

- **Sol-gel**: Chemical process in wet phase starts from the synthesis of a precursor (alkoxide, salt, or other inorganic chemical compounds) with alcohol or water under mild thermal conditions. With the help of a catalyst (base or acid), it is hydrolyzed, forming a hydroxide of the metal used, thus constituting the "sol." Subsequently the groups generated condense, obtaining a three-dimensional structure that forms the "gel" of the metallic oxide. During both processes the oxidation and reduction intervene and these in turn in its aging [18].

Figure 4 shows the diagram of the products and sizes that can be obtained in each of the stages of the sol-gel method.

3. Biomaterial characterization

Through the characterization it is possible to define the structure of biopolymers and materials, with the purpose of explaining and relating the behavior of these with their properties dependent on their structure. Once a polymer and its known properties have been characterized, it is possible to optimize and control the microscopic factors including chemical composition, molecular size, topology, microstructure, the morphology of the aggregates, and transition structure affect the properties of a biopolymer.

From the various tools that enable the characterization of biopolymers, it is possible to classify or divide them according to spectroscopic, physical, mechanical, thermal, and physicochemical properties.

Due to the composition and characteristics of biopolymers, some tools become more related to a certain group of materials; that is why the following are the most commonly used tools for most biopolymers, in order to show an overview general.

3.1 UV-Vis spectroscopy

This technique is very used for quantitative analysis; however, in the qualitative analysis, in the determination of structures; it is surpassed by other techniques, such as infrared spectroscopy and nuclear magnetic resonance. Ultraviolet and visible radiation is characterized by being absorbed by valence electrons of molecules and atoms, which are excited at higher energy levels.

The absorption of electromagnetic radiation by valence electrons is normally within the ultraviolet region of the spectrum; this means that, commonly, matter is opaque to radiation somewhere in this region. In the case of electrons that participate in double bonds, the characteristic absorption frequencies can extend in the visible region, originating color in many organic substances, and in special cases in the near-infrared region [20].

3.2 Infrared spectroscopy (IR)

The vibrational analysis of polymeric materials through Raman and infrared spectroscopy is an appropriate experimental method to obtain information on structural parameters of the same. Thus, besides being able to analyze the chemical species present in the compound, it is possible to obtain, among others, data on the state of order of the polymers (chain orientation, crystallinity, crystalline phases, etc.).

This method is fast and sensitive and does not present great difficulty of interpretation. It is based on the vibration of the atoms of an organic molecule due to thermal energy. Each molecule has a resonant point, analogous to the resonance vibration of mechanical structures. Therefore, the electromagnetic radiation incident on a material is absorbed only in frequencies corresponding to molecular vibrations, if the intensity of radiation-transmitted frequency plotted against the absorption bands of the material (absorption spectrum) is obtained. IR spectroscopy allows to measure the vibrational energy levels of the molecules [21]. Because the levels of vibrational energy are different for each molecule (as well as its isomers), the IR spectrum can be considered as a fingerprint of each molecule.

The parameters of the characteristic bands, measured in IR spectroscopy, are frequencies (energy), intensity (polar character), shape of the band, and polarization in several ways, that is, the transition moment direction of the molecular system. Since the levels of vibrational energy are different for each molecule (and its isomers), the IR spectrum can be considered as the fingerprint of the molecule. However, the identification procedures are based on the purity of the compounds; therefore, it is necessary to verify the purity prior to an IR spectrum analysis.

The interpretation of IR spectra, the wave number ν, which is the number of waves per centimeter, is commonly used; the ratio between ν and ν and the wavelength λ is defined by the following equation:

$$\nu(cm^{-1}) = \frac{10^4}{\lambda}(mm) \quad (1)$$

Group	Frequency range (1/cm)
Stretch vibrations CH	
=CH	3280–3340
=CH	3000–3100
C—CH_3	2872.2962 (+/−10)
O—CH_3	2815–2832
N—CH_3 (aromatic)	2810–2820
N—CH_3 (aliphatic)	2780–805
CH_2	2853–2926 (+/−10)
CH	2880–2900
Stretching vibrations C=O	
Not conjugated	1700–1900
Conjugate	1590–1750
Amides	1650
CH flexing vibrations	
CH_2	1405–1465
CH_3	1355–1395, 1430–1470
Source: *Serrano and Mendizábal [22].*	

Table 1.
Characteristic wave numbers for some functional groups.

The said formula can also be expressed as follows:

$$\nu(cm^{-1}) = 3x10^{10}\,Hz \tag{2}$$

The wavenumber scale is directly proportional to the energy and absorbed vibrational frequency, which corresponds to the positions of the characteristic infrared bands of some functional groups in polymer chains. **Table 1** presents characteristic wave numbers for some functional groups [22].

3.3 Raman spectroscopy

This is a very useful technique to identify chemical compounds. Their results are equivalent to the fingerprint of the compound to be identified. Unlike most other analytical techniques, a chemical or physical pretreatment is not necessary to obtain a Raman spectrum. Hydroxyl groups and silicates have a weak Raman dispersion which means that water and glass do not influence the spectrum obtained. In this technique, the change in wavelength is observed because the molecule disperses the incident radiation inelastically. The gain or loss of energy due to dispersion represents the energetic differences between the vibrational and rotational states of the molecules. This interaction depends on the nature of the polarization ellipsoid of the molecule and the electric dipole moment [23].

Generally is preferable to use Raman spectroscopy to characterize the polymers by the bands associated with the vibrations of the polymer chain which are more intense in the Raman spectrum than in the IR spectrum. The polymers and their reaction mechanisms can be characterized using Raman spectroscopy, from which qualitative and quantitative information is obtained such as stereoregularity, chemical nature, orientation, conformation, and three-dimensional order in the polymer [24].

3.4 Differential scanning calorimetry (DSC)

Calorimetry allows to see very subtle changes in the structure of biopolymers compared to other materials when they are subjected to a processing with elevated temperatures, but that does not undergo any transition; that is why the physical transition temperatures are important in the characterization of biopolymers.

Differential scanning calorimetry (DSC) determines the amount of heat required to maintain the temperature of the sample at a value given by the temperature program. The said technique is measured by determining the heat of the sample through an external thermocouple. In turn measurements are compared to a reference material with a known specific heat; also the specific heat of the sample is determined by comparing the reading obtained from the instrument, corrected with the target, at a constant temperature, and then obtained at a constant rate of heating or cooling. It is said that the glass transition temperature occurs when the movement of the polymer segments begins [21].

3.5 Light scattering

The dispersion density and refractive index of light appear dissolutions and mixtures of liquids, that due to fluctuations in composition [21]. The calculation is determined by Debye, with which the effect of these fluctuations is obtained by relating them to the change in concentration c associated with the osmotic pressure (π) per mole of solute, the turbidity (τ—Greek letter "tau") You can describe how:

$$\tau = \frac{32\pi^3 RTc}{3\lambda^4 N_0} \; n\frac{dn}{dc} \; / \; \frac{d\pi}{dc} \tag{3}$$

where n is the refractive index, λ is the wavelength, R is the universal constant of gases, N_0 is the number of Avogadro; and T the absolute temperature. Inserting the relationship between osmotic pressure and molecular weight gives the Debye equation:

$$K\frac{c}{R_{90}} = H\frac{c}{\tau} = \frac{1}{M} + 2A_2c + \ldots$$

$$K = \frac{2\pi^2 n^2}{N_0\lambda^4}\left(\frac{dn}{dc}\right)^2 \tag{4}$$

$$H = \frac{32\pi^3 n^2}{3N_0\lambda^4}\left(\frac{dn}{dc}\right)^2$$

where M is the molecular weight and A_2 is the second coefficient of the virial. This equation lays the basis for the determination of molecular weights by scattering light.

3.6 Ebullioscopy and cryoscopy

These methods based on Raoult's law depend on the sensitivity of the available thermometry [22]. The average numeral molecular weight is based on the Clausius-Clapeyron equation using the boiling point elevation and the freezing point decrease, as shown below:

$$M_n = \frac{RT^2V}{\Delta H} \frac{C}{\Delta H}_{C \to 0} \quad (5)$$

3.7 Osmometry

The osmotic pressure (π) of closely colligative properties therefore depends on the number of particles, measuring the osmometric pressure p applied to the determination of the osmotic pressure of solvent relative to polymer solutions [21]. An automatic membrane osmometer measures the non-limited capillary rise of a polymer solution using the modified Van't Hoff equation:

$$\pi = \frac{RT}{M_n}C + BC^2 \quad (6)$$

According to **Figure 5**, the inverse of the arithmetic mean of the molecular weight ($M_n - 1$) is the cutoff of the curve π/cRT as a function of c when extrapolated to zero concentration.

3.8 Size exclusion chromatography

This technique can be used to determine the molecular weight by means of polymeric analytes, such as natural molecules (polysaccharides, starches, etc.) and synthetic polymers (polyethylene glycol or polyethylene). To obtain information on the distribution of molecular weights in polydispersed polymers, there is a specific software. It requires a selection of appropriate columns for a correct analysis. Columns packed with polymeric absorbers are often used for polymer molecules with a wide molecular weight distribution, such as heparin, starch, or cellulose [25].

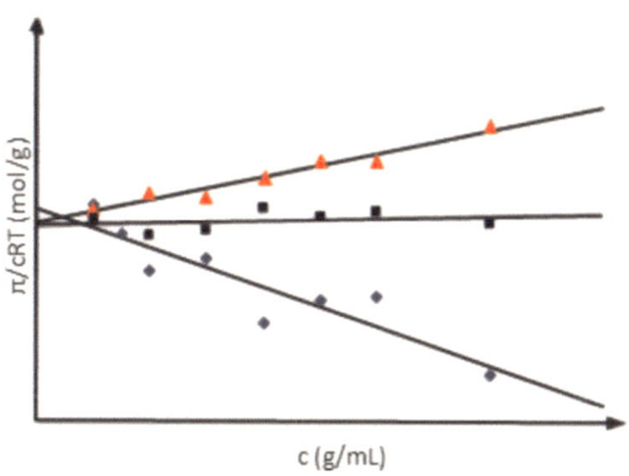

Figure 5.
Obtaining M_n through osmometry, the value at zero concentration is extrapolated from the experimental measurements.

3.9 Scanning electron microscopy (SEM)

The electronic scanning microscopy (SEM) allows obtaining characteristics of the surface of materials, as well as the shape and size of their particles and their arrangement. The operation of this technique is that the electrons travel through the arrangement of lenses designed to obtain a convergent beam. The coils located under the array of lenses direct the beam of electrons from left to right and from top to bottom in such a way that a sweep is made on the entire surface of the sample at the base of the vacuum chamber. The electrons that hit the sample are diffracted to the detector. The latter captures the signal and sends it to a processor 10 where it is converted into an image [26].

3.10 X-ray diffraction (XRD)

The depth of penetration of laser radiation in the visible region is only several hundred Angstroms, so Raman spectroscopy is an excellent method to perform a structural description of surface areas of the compounds and a good complement to the structural information obtained by X-ray diffraction; this allows to make a spectroscopic map of the surface of the compounds. Each component has its characteristic signal region; therefore, X-ray diffraction works as a fingerprint identification. In the same way, it is possible to determine in addition to the morphology of biopolymers its stability, observing changes in the ranges and intensity of the signals with respect to time, using this technique for periods after the analysis in an initial time [27].

3.11 Thermogravimetric analysis (TGA)

Thermogravimetry is used in the study of primary reactions related to the decomposition of solid and liquid materials; this technique allows analyzes *r* desorption processes and adsorption and decomposition reactions in an atmosphere of inert gas or in the presence of oxygen [28]. However, this technique does not allow knowing the chemical composition of the material under study or identifying the thermal changes that are not associated with mass variations such as crystallization or glass transition. Basically, a thermogravimetric analysis consists of the continuous recording of the variation of the mass of the material according to the variation of the temperature at a constant thermal rate. This type of thermogravimetric process is known as dynamic analysis. There is the option of doing an isothermal thermogravimetric analysis, in which the constant temperature is maintained for a set period. As a result of the thermogravimetric analysis, the mass change data are obtained with respect to temperature or time and a thermogram, which graphically represents the percentage variations of the mass. This type of technique is widely used in the quantitative characterization and kinetic characterization of polymers, coal, and clays, among other materials. Even in Costa Rica, this technique is applied to the analysis of soils, food products, and crops, among other areas [29].

4. Conclusion

At the moment countless equipment have been generated that allow the encapsulation of different organic and inorganic materials; however, there are different basic tools for the development of new materials. However, important factor for the development of these structures is the method used with which we can manipulate its shape and size through the study of the different variables that constitute it with the aim of obtaining a material that meets the needs to eradicate their problem.

Author details

F.F. Razura-Carmona[1], G.A. Prado-Guzmán[1], A. Perez-Larios[2], M.V. Ramírez-Marez[3], M. Herrera-Martínez[4] and Jorge Alberto Sánchez-Burgos[1]*

1 División de Estudios de Posgrado e Investigación, Tecnológico Nacional de México/Instituto Tecnológico de Tepic, Tepic, Nayarit, Mexico

2 División de Ciencias Agropecuarias e Ingenierías, Centro Universitario de los Altos, Universidad de Guadalajara, Tepatitlán de Morelos, Jalisco, Mexico

3 Departamento de Ingenierías Química y Bioquímica, Tecnológico Nacional de México/Instituto Tecnológico de Morelia, Morelia, Mexico

4 Instituto de Farmacobiología, Universidad de la Cañada, Teotitlán de Flores Magón, Oax., Mexico

*Address all correspondence to: jsanchezb@ittepic.edu.mx

References

[1] Shen Y. Rice husk silica derived nanomaterials for sustainable applications. Renewable and Sustainable Energy Reviews. 2017;**80**:453-466

[2] Gu M, Zhang Q, Lamon S. Nanomaterials for optical data storage. Nature Reviews Materials. 2016;**1**(12):1-14

[3] Khademolhoseini S, Talebi R. Green synthesis and characterization of cobalt aluminate nanoparticles and its photocatalyst application. Journal of Materials Science: Materials in Electronics. 2016;**27**(3):2938-2943

[4] Daniel MC, Astruc D. Gold nanoparticles: Assembly, supramolecular chemistry, quantum-size-related properties, and applications toward biology, catalysis and nano technology. Chemical Reviews. 2004; **104**:293-346

[5] Medintz IL, Uyeda HT, Goldman ER, Mattoussi H. Quantum dot bioconjugates for imaging, labelling and sensing. Nature Materials. 2005;**4**: 435-446

[6] Lu AH, Salabas EL, Schuth F. Magnetic nanoparticles: Synthesis, protection, functionalization, and applications. Angewandte Chemie International Edition. 2007;**46**: 1222-1244

[7] Mohammad R, Navid R, Reza S, Ghazal R. Introduction to Nanomaterials in Medicine, Chapter 1. IOP Concise Physics. Morgan & Claypool Publishers; 2019. pp. 1-4

[8] Conesa-Egea J, Zamora F, Amo-Ochoa P. Perspectives of the smart Cu-iodine coordination polymers: A portage to the world of new nanomaterials and composites. Coordination Chemistry Reviews. 2019;**381**:65-78

[9] Wang X, Gong Y, Shi G, Leong W, Keyshar K, Ye G, et al. Chemical vapor deposition growth of crystalline monolayer MoSe2. ACS Nano. 2014; **8**(5):5125-5131

[10] Rajput N. Methods of preparation de nanoparticles. International Journal of Advances on Engineering & Technology. 2015;7(4):1806-1811

[11] Zhu H, Zhang C, Yin Y. Rapid synthesis of copper nanoparticles by sodium hypophosphite reduction in ethylene glycol under microwaves irradiation. Journal of Crystal Growth. 2004;**270**:722-728

[12] de-Jong KP. Synthesis of Solid Catalysts. Wiley-VCH: Weinheim; 2009

[13] Schmid G. Nanoparticles. From Theory to Application. Weinheim: Wiley-VCH; 2004

[14] Suresh S, Arunseshan C. Dielectric properties of cadmium selenide (CdSe) nanoparticles synthesized by solvothermal method. Applied Nanoscience. 2014;**4**(2):179-184

[15] Karakhanov E, Maximov AL, Zakharyan EM, Zolotukhina AV, Ivanov AO. Palladium nanoparticles on dendrimer-containing supports as catalysts for hydrogenation of unsaturated hydrocarbons. Molecular Catalysis. 2016;**440**:107-119

[16] Kim Y-G, Oh S k, Crooks RM. Preparation and characterization of 1–2 nm dendrimer-encapsulated gold nanoparticles having very narrow size distributions. Chemistry of Materials. 2004;**16**(1):167-172

[17] Zheng J, Dickson RM. Individual water-soluble dendrimer-encapsulated silver nanodot

fluorescence. Journal of the American Chemical Society. 2002;**124**: 13982-13983

[18] Ashour AH, El-Batal AI, Maksoud MI, Sayyad GS, Labib S, Abdeltwab E, et al. Antimicrobial activity of metal-substituted cobalt ferrite nanoparticles synthesized by sol-gel technique. Particuology. 2018;**40**: 141-151

[19] Seidel A. Characterization and Analysis of Polymers. John Waley & Sons, Inc, Publication. 2008. pp 1-31

[20] Malacara D. Óptica básica. Segunda edición ed. Mexico, FCE-SEP; 2004. (Ciencia y Tecnología)

[21] Campbell D, Pethrick RA, White JR. X-ray diffraction. In: Polymer Characterization: Physical Techniques. UK: Stanley Thornes; 2000. pp. 194-236

[22] Serrano RFL, Mendizábal ME. Introducción a la ciencia de los polímeros. Primera Edición ed. Guadalajara, México: Universidad de Guadalajara; 2015

[23] Frausto-Reyes C, Medina-Gutiérrez C, Sato-Berrú R, Saghún LR. Qualitative study of ethanol content in tequilas by Raman spectroscopy and principal component analysis. Spectrochimica Acta Part A. 2005;**61**: 2657

[24] Rudin A, Choi P. The Elements of Polymer Science and Engineering. 3rd ed. Kidlington, Oxford, UK: Elsevier; 2012. pp 149-228

[25] Gutiérrez-Bouzán MC, Burdó-Expósito A, Cegarra-Sánchez J. La cromatografía de exclusión: análisis de la distribución de pesos moleculares en siliconas por GPC. Boletín Intexter del Instituto de Investigación Textil y de Cooperación Industrial. 2009;**135**(1): 33-40

[26] Valero-Valdivieso M, Ortegón Y, Uscategui Y. Biopolimeros: Avances y Perspectivas. DYNA. 2013;**80**(181): 171-180

[27] Ezquerra TA, Martínez-Gómez A, Álvarez C, Alonso E, Sanz A, García-Gutiérrez MC, et al. Structure-dynamics relationship during the amorphous to smectic transition of a main chain liquid crystalline polymer. Journal of Non-Crystalline Solids. 2005;**351**(33–36): 2768-2772

[28] Grueiro LF. Estudio cinético, dinamomecánico y termogravimétrico del sistema epoxídico BADGE (n=O)/m-XDA mediante las técnicas de análisis térmico: DSC, DMA y TGA, construcción de un diagrama TTT [doctoral dissertation]. Universidade de Santiago de Compostela; 2001

[29] Rodríguez E, Villegas E. Caracterización de polímeros aplicando el método termogravimétrico. Métodos y materiales. 2012;**2**(1):25-32

Chapter 7

Phosphate-Mineralization Microbe Repairs Heavy Metal Ions That Formed Nanomaterials in Soil and Water

Xiaoniu Yu and Qiwei Zhan

Abstract

This chapter presents a new method for treatment of heavy metal ions in soil or water. Heavy metal pollution in soil and water has become one of the serious environmental problems. Heavy metal pollution can degrade soil quality and ecosystems, contaminate crops, and threaten human health. At present, there are three ways to repair heavy metals in soil or water, including physical, chemical, and biological technologies. The microbial mineralization technology can be applied to remove heavy metal pollutants which contaminated soil and water and has been paid with more attention in recent years. Heavy metal ions can be mineralized by phosphate-mineralization microbe to form stable phosphate nanomaterials compared to mineralization of carbonate-mineralization microbe in the environments. Therefore, heavy metal pollution can well be removed from soil or water by microbial mineralization method.

Keywords: heavy metal pollution, soil, water, phosphate-mineralization microbe, phosphate nanomaterials

1. Introduction

The density of the heavy metal is bigger than 4.5 g/cm^3 or more, and atomic numbers are from 23 (V) to 92 (U) of heavy metal elements, such as lead, nickel, zinc, copper, iron, cadmium, chromium, etc. [1]. They are widely used in industrial production and discharged into the environment due to the failure to process heavy metals in mining and industrial production. Soil and water are the ultimate destination of these heavy metals due to their particularity. Compared with air pollution, water pollution, and industrial solid waste pollution, heavy metals in soil and water are invisible and concealed. The pollution of heavy metals in soil and water can lead to the degradation of soil fertility, the reduction of crop yields, and the decline in quality, which seriously affects the environmental quality and sustainable economic development and threatens people's food safety [2]. Heavy metal pollution has become a global concern [1, 3–9]. Whether in the soil or water, different heavy metals enter the bottom of the human food chain and finally enter the human body [2]. Heavy metals are very difficult to be biodegraded and can easily be biomagnified in the human body. Heavy metals can interact strongly with proteins and enzymes in the human body, making them inactive, or they can accumulate in certain organs of the human body, causing chronic poisoning, which is a serious threat to people's lives, health, and safety [2, 5].

The content, migration, and damage of heavy metal ions are high in the electroplating wastewater irrigation area, the tailings accumulation area, and the surrounding cultivated soil. Since heavy metal pollution has already caused a major threat to human survival and health, countries around the world have formulated corresponding laws and regulations to strictly limit the emission of heavy metals. At the same time, areas of heavy metal pollution can well be repaired through physical, chemical, and biological technologies to reduce the harm [9]. Physical and chemical methods to repair heavy metal pollution have been widely reported, including chemical curing, soil leaching, electrokinetic repair, etc. The physicochemical methods may cause secondary pollution because of its damage to soil structure and indigenous microorganisms. Bioremediation technology is a more effective method for controlling polluted soil and water and has become a new hotspot in current environmental engineering science and technology research, including phytoremediation, microbial adsorption, and microbial mineralization technologies. The use of microorganisms (bacteria, algae, yeast, etc.) to reduce or eliminate heavy metal pollution has been reported widely at home and abroad. The mechanism of bioremediation heavy metals includes: (1) cell metabolism—a specific metabolic pathway can make heavy metal ions precipitation or into low-toxic; (2) biosorption—using living cells, inanimate organisms metal-binding proteins and polypeptides or biopolymers as biosorbents, vacuolar swallowing, precipitation and redox reactions; and (3) biomineralization—heavy metal ions are precipitated into biominerals by microbial hydrolysate.

At present, the problem of heavy metal pollution in soil and water is becoming more and more serious, and the use of carbonate-mineralization microbe to repair heavy metals pollution has attracted people's attention because of its excellent characteristics [10, 11]. Although many studies have been carried out, the stability of carbonate in the environment is lower than that of phosphate. Therefore, this chapter proposes that phosphate-mineralization microbe mineralizes heavy metal ions to form stable phosphates. Based on the principle of biomineralization, organic phosphate monoester is hydrolyzed by phosphatase produced by phosphate-mineralization microbe of growth and metabolism and then obtains phosphate ions. Heavy metal ions can combine with phosphate ions to form phosphate minerals in soil or water, thereby reducing the risk of heavy metals being absorbed by plants and the harm to the natural environment.

2. Methodology

Bacillus subtilis was purchased by China Center of Industrial Culture Collection. Pure water was made by small ultrapure water machine. All raw materials were of analytically pure grade and used without further purification.

Bacillus subtilis of cultivation [12, 13]: 5 g of tryptone and 3 g of beef extract were completely dissolved in 1 L of pure water, and its pH was adjusted to 7.0 using diluted NaOH solution. 1 L of above solution was added to two beaker flasks (1000 mL), and each bottle was added 500 mL of mixture solution and wrapped by paper and gauze. Two beaker flasks were placed in autoclave under 125°C and 0.1 Mpa for 30 min. Then they were cooled to room temperature under natural conditions. Next, 3 mL of *Bacillus subtilis* strains were added to each beaker flask and cultivated in the oscillation incubator ($170 r\cdot min^{-1}$) at 28°C for 22–24 hours. As a result, the harvested microorganisms were kept in a refrigerator at 4°C before use. Optical density (OD) of *Bacillus subtilis* liquid was measured by UV spectrophotometer at 600 nm of wavelength with OD range of 1.23–1.36.

Syntheses of barium phosphates [14]: organic phosphate monoester (pH = 8.8–9.3) was added to 100 mL of *Bacillus subtilis* under static conditions for 24 hours at room temperature. The mixture solution of *Bacillus subtilis* and organic phosphate monoester was centrifuged, and then the centrifugal solution was obtained. The pH of solution was adjusted to 7, 8, 9, 10, and 11 with diluted HCl and NaOH solution. $BaCl_2{\cdot}2H_2O$ powder was added to the above solution at different pH. Precipitated solution was stilled for 24 hours at room temperature after stirring for 1 min. All products were dried at 60°C for 24 hours. After that, all powder precipitates were collected.

Microbiological reduction-precipitation of cerium phosphates [15]: Organic phosphate monoester (pH = 8.8–9.3) was added to 100 mL of *Bacillus subtilis* under static conditions for 24 hours. Next, 0.01 and 0.02 mol of $Ce(SO_4)_2$ were added to above solution. The precipitated solution was stood under static conditions for 24 hours. As a result, two powder precipitates were dried at 60°C for 24 hours. After that, two powder specimens were collected and characterized.

Syntheses of iron phosphates [16]: Organic phosphate monoester (pH = 8.8 9.3) was injected into 200 mL of *Bacillus subtilis* and was allowed to stand under static conditions for 24 hours at room temperature. After 24 hours, $FeSO_4{\cdot}7H_2O$ (12 mM) was added to the above mixture solution with 2 min of stirring. The precipitated solution was allowed to stand under static conditions for 72 hours at room temperature. The products were then dried at 60°C for 24 hours. After that, powder precipitates were obtained and characterized.

Fourier-transform infrared spectroscopy (FTIR) spectra of the samples were recorded with a Nicolet 5700 spectrometer using the KBr pellet technique in the range of 399–4000 cm^{-1}. The phase purity of products was examined by powder X-ray diffraction (XRD) with a Bruker D8-Discover diffractometer with Cu Kα radiation (λ = 1.5406 Å). Scanning electron microscopy (SEM, FEI Company, the Netherlands, operating voltage 20 kV) with an energy-dispersive X-ray spectroscopy (EDS) was used to conduct morphological studies and to measure elemental compositions of the samples. Transmission electron microscopy (TEM) images were obtained on a FEI, G2 20 equipment. The samples were dispersed in anhydrous ethanol before being tested. Thermogravimetric-differential scanning calorimetry (DSC-TG) was carried out on STA449 F3 thermogravimetric analyzer (NETZSCH, Germany). The analyses were carried out simultaneously in a nitrogen atmosphere at a heating rate of 10°C/minute between room temperature and 700–800°C under an N_2 flow of 20 $mL{\cdot}min^{-1}$. The average crystallite size of biophosphate nanomaterials was estimated by the Scherrer formula [13]:

$$D = \frac{k\lambda}{\beta \cos\theta} \tag{1}$$

where D is the average crystallite size of samples (nm), k is the shape factor (0.9), λ is the wavelength of Cu Kα radiation equal to 1.5406 Å, β is defined as the full width of the peak at half of the maximum intensity (FWHM), and θ is the diffraction angle of the selected reflection.

3. Barium phosphate nanoparticles prepared by microbial mineralization method

The FTIR spectra of barium phosphates obtained at pH = 8 and 11 were displayed in **Figure 1**. The peak at 3415.62 cm^{-1} was KBr mull absorption, which appears in the spectrum of any compounds. A shoulder peak at 2973.47 cm^{-1}

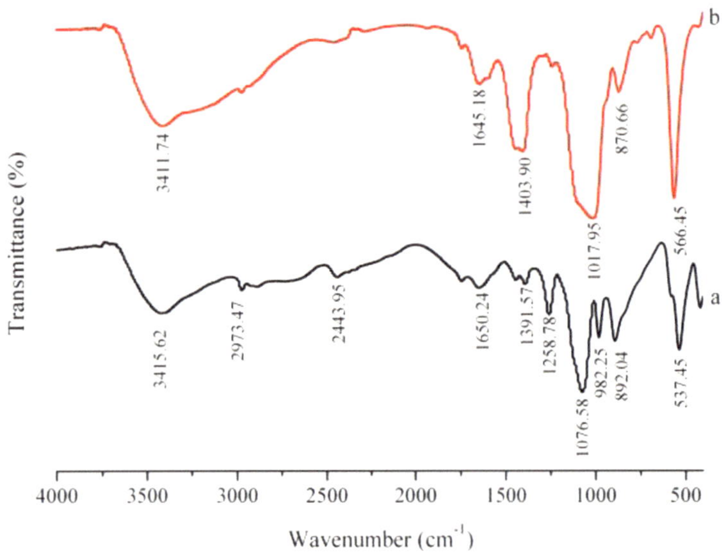

Figure 1.
FTIR spectra of barium phosphates prepared at pH = 8 (a) and 11 (b) [14].

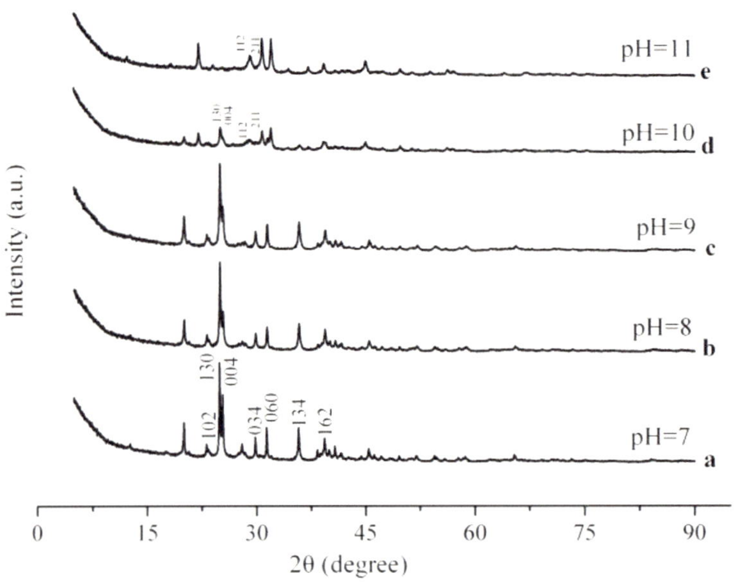

Figure 2.
XRD patterns of barium phosphates precipitated at pH = 7, 8, 9, 10, and 11 [14].

corresponded to hinder rotation. The peak at 1650.24 cm^{-1} reveals thermal vibrations of H^+ in an intermolecular phonon vibration associated with H^+-PO_4^{3-}. The peak at 1258.78 cm^{-1} was P=O stretching vibration. The peaks at 1391.75, 1076.58, and 982.25 cm^{-1} corresponded to stretching vibration of P-O (H). The P=O of deformation vibration was found at 892.04 cm^{-1} and to be nondegenerate. The P-O(H) wagging and rocking vibration was found at 537.45 cm^{-1}. The above results are similar with literatures reported, as shown in **Figure 1a** [17–20]. The peaks are at 3411.74, 1645.18, 1403.90, 1017.95, 870.66, and 566.45 cm^{-1} in the FTIR spectrum with respect to the OH and PO_4^{3-} (**Figure 1b**), which were in accordance with the data reported in the literature [21].

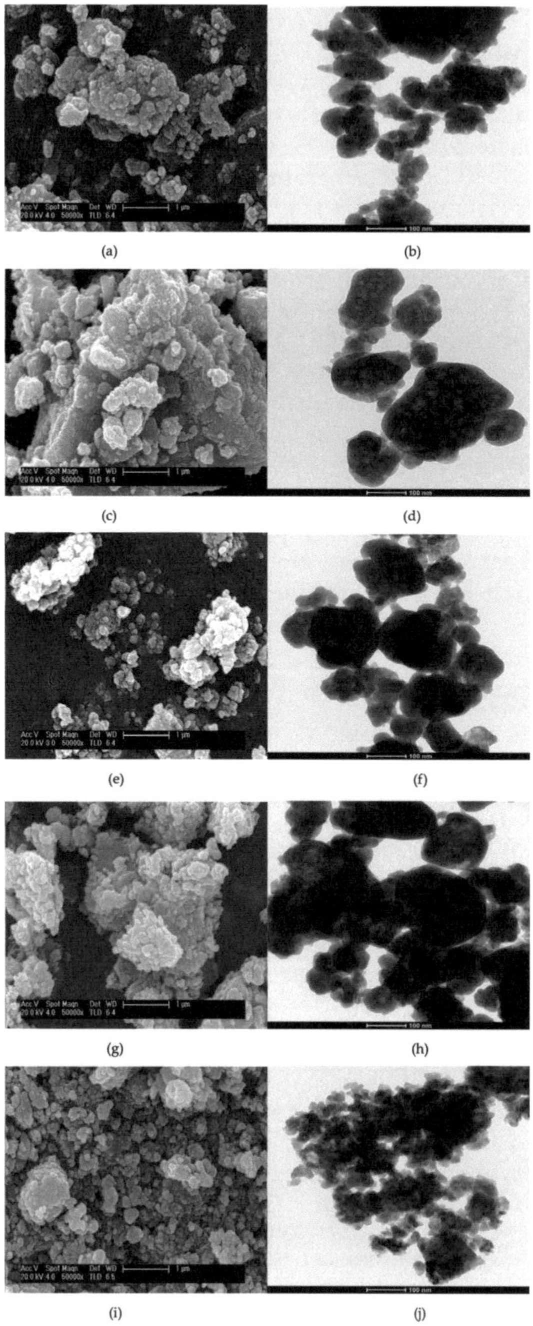

Figure 3.
SEM and TEM images of barium phosphates: (a, b) pH = 7, (c, d) pH = 8, (e, f) pH = 9, (g, h) pH = 10, and (i, j) pH =11 [14].

Chemical compositions of barium phosphates were analyzed by XRD technique, as shown in **Figure 2**. Barium phosphates of patterns could be readily indexed using the reported structures of $BaHPO_4$ (JCPDS Card No. 72–1370) at pH < 10 (**Figure 2a–c**). When pH was 10, the mixture of $BaNaPO_4$, $BaHPO_4$, and $Ba_5(PO_4)_3OH$ (JCPDS No. 70–1787, 09–0113 and 36–0272) was found, as shown in **Figure 2d**. The mixture of $BaNaPO_4$ (JCPDS No. 70–1787) and $Ba_5(PO_4)_3OH$ (JCPDS No. 78–1441) was precipitated at pH = 11(**Figure 2e**). Above results indicated that HPO_4^{2-} and PO_4^{3-} ions were obtained by *Bacillus subtilis* hydrolyzing organic phosphate monoester. HPO_4^{2-} ions were formed at pH = 7, 8, and 9. HPO_4^{2-} and PO_4^{3-} ions were formed at pH = 10. PO_4^{3-} ions were formed at pH = 11. Therefore, the barium phosphates were mainly $BaHPO_4$ at pH < 10 [22]. The barium phosphates were mainly the mixture of $BaNaPO_4$, $BaHPO_4$, and $Ba_5(PO_4)_3OH$ at pH = 10 [22]. The barium phosphates were mainly the mixture of $BaNaPO_4$ and $Ba_5(PO_4)_3OH$ at pH = 11 [22]. Organic phosphate monoesters can well be hydrolyzed through *Bacillus subtilis*.

SEM and TEM images showed shape of barium phosphates were nanoparticles, as shown in **Figure 3**. SEM images showed that barium phosphates were nano-agglomerates, (**Figure 3a, c, e, g**, and **i**). The size of the nano-agglomerates was less than 3 μm. TEM images indicated that morphology of barium phosphate particles was mainly irregular flakes with diameter in the range of 20–100 nm, as shown in **Figure 3b, d, f, h**, and **j**. The average size of barium phosphates was calculated by the Scherrer formula (1), and the average size was 48.56, 36.60, 31.89, 33.27, and 34.79 nm when pH were 7, 8, 9, 10, and 11, respectively [13]. Above results showed that barium phosphate nanomaterials could be easily prepared by the centrifuged solution of *Bacillus subtilis* and organic phosphate monoester reacting with $BaCl_2{\cdot}2H_2O$, which were different from chemical and biomineralization methods [17, 22].

4. Microbiological reduction-precipitation of nanostructured cerium phosphates

Elemental compositions of cerium phosphates were analyzed by EDS spectra, as shown in **Figure 4**. The elements C, O, Na, P, S, and Ce were found in cerium phosphates when P/Ce molar ratio was 1:1 and 2:1 (**Figure 4a** and **b**). The raw files and cerium phosphates could be indexed by MDI jade 5.0 procedure. The indexed results indicated $CePO_4@NaCe(SO_4)_2(H_2O)$ and $CePO_4$ were found when P/Ce molar ratio was 1:1 and 2:1, respectively, as shown in **Figure 5a** and **b**. The standard cards of $CePO_4@NaCe(SO_4)_2(H_2O)$ used were JCPDS Card No. 74–1889 and 86–0526. $CePO_4$ was JCPDS Card No. 75–1880.

The average crystallite size of cerium phosphates was calculated via the Scherrer formula (1), as shown in **Table 1** [13]. The results showed that the average size of $CePO_4@NaCe(SO_4)_2(H_2O)$ nanoparticles was bigger than $CePO_4$.

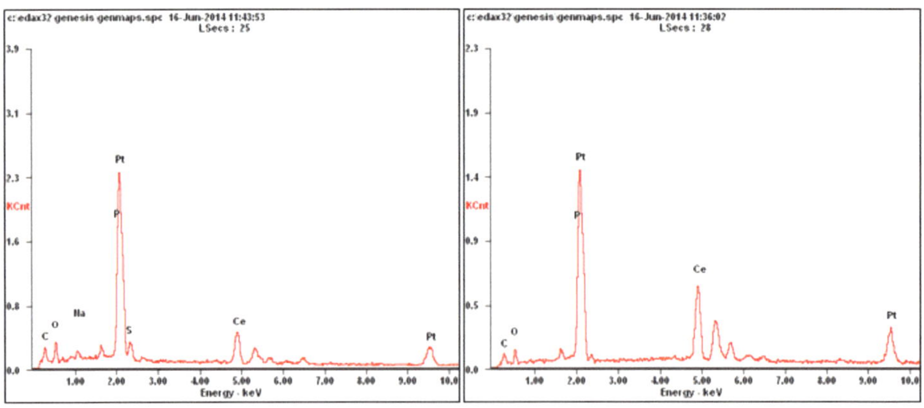

(a) (b)

Figure 4.
EDS spectra: (a) $CePO_4@NaCe(SO_4)_2(H_2O)$ and (b) $CePO_4$ [15].

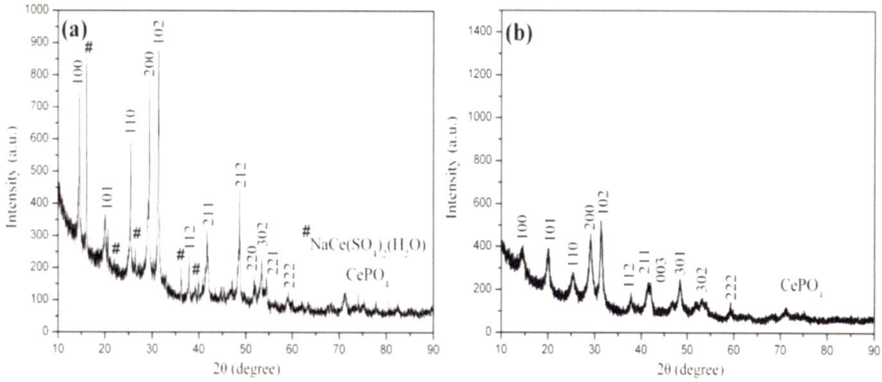

Figure 5.
XRD patterns: (a) $CePO_4@NaCe(SO_4)_2(H_2O)$ and (b) $CePO_4$ [15].

Sample	P/Ce (molar ratio)	θ (deg)	β (deg)	D (nm)
$CePO_4$	2:1	15.67	0.504	15.61
CePO4@NaCe(SO4)2(H2O)	1:1	15.68	0.243	32.34

Table 1.
Average crystallite size of cerium phosphates [15].

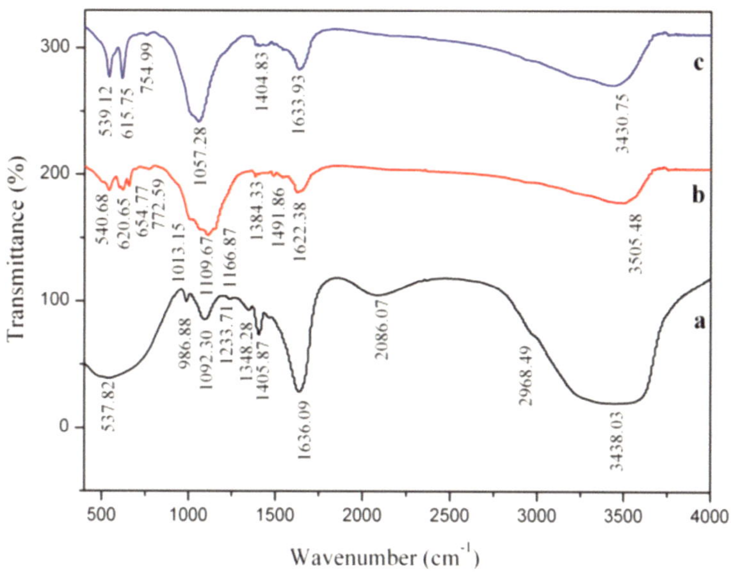

Figure 6.
FTIR spectra: (a) Bacillus subtilis, (b) $CePO_4@NaCe(SO_4)_2(H_2O)$, and (c) $CePO_4$ [15].

Cerium phosphates and *Bacillus subtilis* were determined by FTIR spectra, as shown in **Figure 6**. According to literature reported, the peaks at 3438.03, 2968.49 and 1348.28, 1636.09, 1233.71, 1092.30, 1405.87, and 537.82 cm^{-1} corresponded to, respectively, OH, C-H, C=O or N-H, C-O, etc. (**Figure 3a**) [23]. Above results showed that these functional groups play important roles in the process of adsorption and reduction of Ce(IV) [23]. **Figure 6b** indicated the peak at 1013.15 cm^{-1} was attributed to the asymmetry stretching vibration of the PO_4^{3-} groups. The peaks

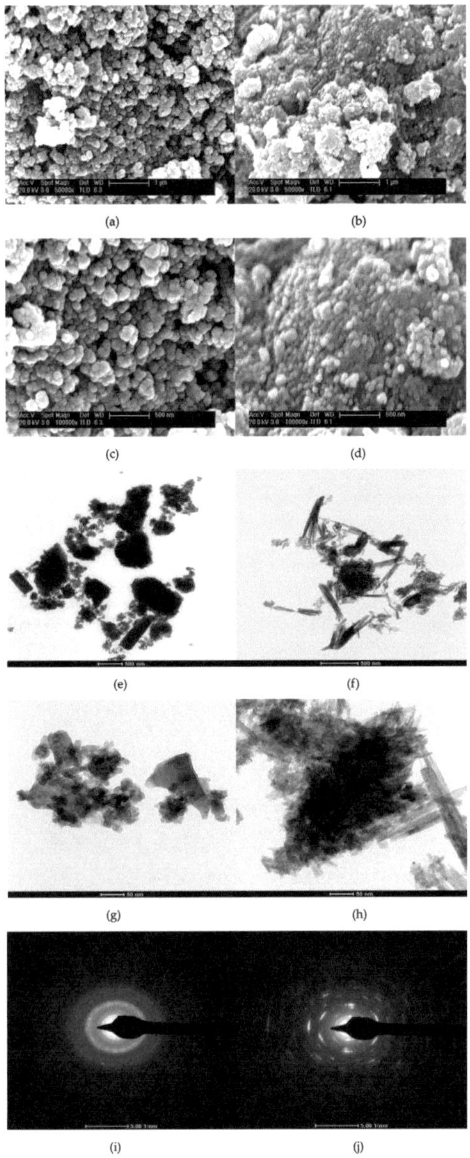

Figure 7.
SEM images of $CePO_4$@$NaCe(SO_4)_2(H_2O)$ (a, c) and $CePO_4$ (b, d), TEM images of $CePO_4$@$NaCe(SO_4)_2(H_2O)$ (e, g) and $CePO_4$ (f, h), and ED images of $CePO_4$@$NaCe(SO_4)_2(H_2O)$ (i) and $CePO_4$ (j) [15].

at 620.65 and 540.68 cm^{-1} were assigned to the O-P-O bending vibration [24]. The peaks at 1109.67 and 654.77 cm^{-1} corresponded to the stretching vibration of SO_4^{2-} group and the bending vibration of O-S-O, respectively. Weak transmission bands of PO_4 were at 1057.28 cm^{-1} and bending at 539.12 and 615.75 cm^{-1}, as shown in **Figure 6c**, in good agreement with the investigations on $CePO_4$ by Yang [24]. Other peaks in **Figure 6b** and **c** may be ascribed to the absorption vibration of N-H, C=O, COO–, etc., according to the FTIR of *Bacillus subtilis*.

The morphology and crystalline nature of cerium phosphates were observed by SEM, TEM, and electron diffraction (ED) images, as shown in **Figure 7**. SEM images showed shape of $CePO_4$@$NaCe(SO_4)_2(H_2O)$ and $CePO_4$ was nano-clusters and sphere-like with a narrow diameter distribution (**Figure 4a, b, c**, and **d**). TEM images further confirmed that a large number of nanoparticles were stacked together and formed the big agglomerates. The size of $CePO_4@NaCe(SO_4)_2(H_2O)$ and $CePO_4$ was in the range of 20–50 nm and 5–25 nm, respectively. Nanosheets and nanorods in $CePO_4@NaCe(SO_4)_2(H_2O)$ and $CePO_4$ were observed, as shown in **Figure 7g** and **h**. ED patterns showed that $CePO_4@NaCe(SO_4)_2(H_2O)$ and $CePO_4$ nanomaterials were well crystallized, as shown **Figure 7i** and **j**.

5. Microbiological reduction-precipitation of nanostructured iron phosphates

FTIR spectrum of iron phosphates was observed in **Figure 8**. The broad peaks at 3000–3500 cm^{-1} and the strong peak at 1644.90 cm^{-1} were attributed to the O-H stretching vibration in water. The iron phosphates showed the peaks at 1056.82 and 553.89 cm^{-1} were stretching vibration of PO_4^{3-} groups [17–20, 21, 24–26]. O, P, and Fe elements were found in EDS spectrum, as shown in **Figure 9a**. Therefore, iron phosphates were composed by O, P, and Fe elements. XRD further confirmed iron phosphates could be readily indexed to the reported structures of $Fe_3(PO_4)_2{\cdot}8H_2O$ (JCPDS Card No.03–0070), and no peaks attributed to impurities (**Figure 9b**).

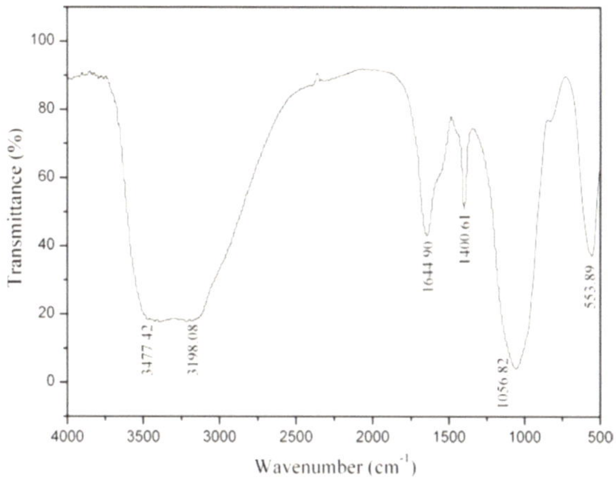

Figure 8.
FTIR spectrum of iron phosphates [16].

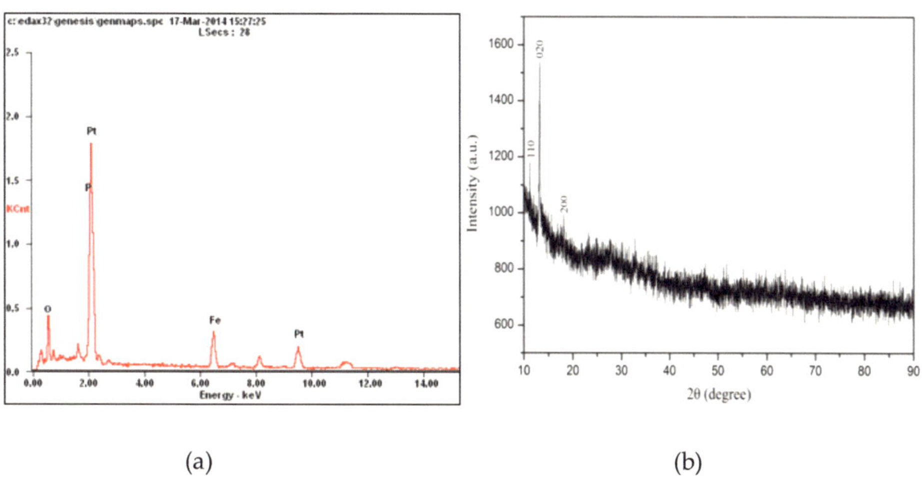

Figure 9.
EDS spectrum and XRD patterns of iron phosphates [16].

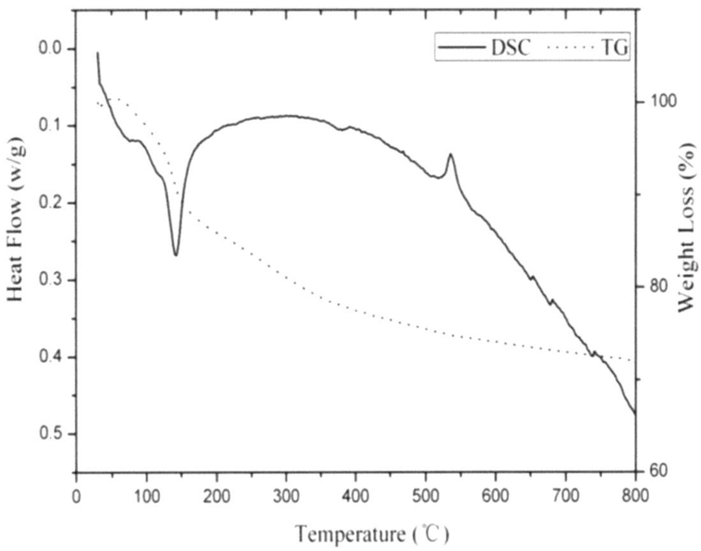

Figure 10.
TG-DSC curves of $Fe_3(PO_4)_2{\cdot}8H_2O$ [16].

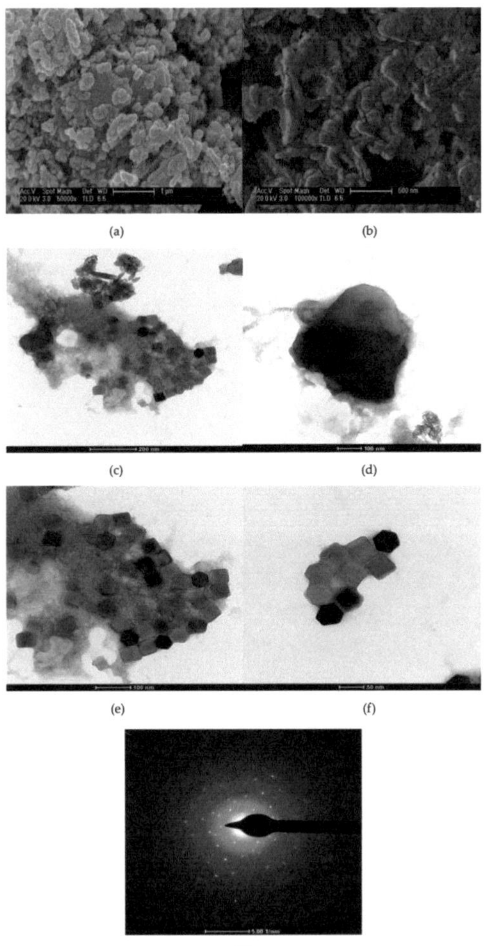

Figure 11.
SEM images (a, b), TEM images (c, d, e, f), and ED patterns (g) of $Fe_3(PO_4)_2{\cdot}8H_2O$ [16].

TG-DSC curves of iron phosphates were analyzed from room temperature to 800°C, as shown in **Figure 10**. The TG curve showed total weight loss was about 28.38%, which consisted with the theoretical value (28.69%). TG also confirmed chemical composition of iron phosphates was $Fe_3(PO_4)_2{\cdot}8H_2O$. Crystal water in $Fe_3(PO_4)_2{\cdot}8H_2O$ could be removed at temperature range of 61–250°C. DSC curve showed endothermic peaks at 142.1°C and 536.1°C were decomposition points of $Fe_3(PO_4)_2{\cdot}8H_2O$ and crystallization of $Fe_3(PO_4)_2$, respectively.

SEM, TEM, and ED images of $Fe_3(PO_4)_2{\cdot}8H_2O$ were observed in **Figure 11**. SEM images showed that the shape of $Fe_3(PO_4)_2{\cdot}8H_2O$ was honeycomb-like, as shown in **Figure 11a** and **b**. TEM images indicated that the morphology of $Fe_3(PO_4)_2{\cdot}8H_2O$ was quadrilateral, hexagonal, and irregular structure with size in the range of 20–200 nm (**Figure 11c, d, e,** and **f**). The average crystallite size was calculated through the Scherrer formula (1), which was 31.89 nm [13]. The ED patterns indicated $Fe_3(PO_4)_2{\cdot}8H_2O$ was polycrystalline in nature (**Figure 11g**).

6. Conclusions

1. *Bacillus subtilis* could hydrolyze organic phosphate monoester to form phosphate ions. Barium phosphate nanoparticles could well be prepared by the centrifuged solution of *Bacillus subtilis* and organic phosphate monoester reacting with barium ions under different pH. TEM images indicated that shape of barium phosphate particles was mainly irregular flakes. The average size of barium phosphates were 48.56, 36.60, 31.89, 33.27, and 34.79 nm when pH were 7, 8, 9, 10, and 11, respectively.

2. Nanostructures of $CePO_4$ and $CePO_4@NaCe(SO_4)_2(H_2O)$ have been successfully synthesized through the mixture solution of *Bacillus subtilis* and organic phosphate monoester reacting with Ce(IV). The average crystalline size of $CePO_4$ and $CePO_4@NaCe(SO_4)_2(H_2O)$ was 15.61 and 32.34 nm, respectively. TEM images indicated that shape of $CePO_4@NaCe(SO_4)_2(H_2O)$ and $CePO_4$ was nanosheets and nanorods with nonuniform size, respectively.

3. FTIR spectrum, EDS, XRD, and TG-DSC confirmed the chemical composition of iron phosphates was $Fe_3(PO_4)_2{\cdot}8H_2O$. TEM images showed that the morphology of $Fe_3(PO_4)_2{\cdot}8H_2O$ powder was quadrilateral, hexagonal, and irregular in structure with the size of 31.89 nm.

4. The content of phosphate ions can be increased by adjusting the amount of substrate and pH in the mixture solution of *Bacillus subtilis* and organic phosphate monoester. At same time, the removal efficiency of heavy metal ions can well be increased in soil and water. The different phosphate nanomaterials can be prepared using centrifugation liquid or bacterial liquid.

Conflict of interest

The authors declare no conflict of interest.

Notes/thanks/other declarations

This work was supported by the National Nature Science Foundation of China (51702238) and the Opening Funds of Jiangsu Key Laboratory of Construction Materials (CM2018–02); the financial support is gratefully acknowledged.

Author details

Xiaoniu Yu[1,2] and Qiwei Zhan[3*]

1 College of Civil Engineering and Architecture, Wenzhou University, Wenzhou, China

2 School of Environment, Tsinghua University, Beijing, China

3 School of Materials Science and Engineering, Anhui University of Technology, Maanshan, China

*Address all correspondence to: zhanqw@ahut.edu.cn

References

[1] Pandey LK, Park J, Son DH, Kim W, Islam MS, Choi S, et al. Assessment of metal contamination in water and sediments from major rivers in South Korea from 2008 to 2015. Science of the Total Environment. 2019;**651**:323-333. DOI: 10.1016/j.scitotenv.2018.09.057

[2] Gardener H, Bowen J, Callan SP. Lead and cadmium contamination in a large sample of United States infant formulas and baby foods. Science of the Total Environment. 2019;**651**:822-827. DOI: 10.1016/j.scitotenv.2018.09.026

[3] Cheng H, Shen RL, Chen YY, Wan QJ, Shi TZ, Wang JJ, et al. Estimating heavy metal concentrations in suburban soils with reflectance spectroscopy. Geoderma. 2019;**336**:59-67. DOI: 10.1016/j.geoderma.2018.08.010

[4] Shi TZ, Guo L, Chen YY, Wang WX, Shi Z, Li QQ, et al. Proximal and remote sensing techniques for mapping of soil contamination with heavy metals. Applied Spectroscopy Reviews. 2018, 2018;**3**:1-23. DOI: 10.1080/05704928.2018.1442346

[5] Peng YT, Shen Y, Ge MY, Pan ZY, Chen WM, Gong B. Efficient extraction of heavy metals from collagens by sulfonated polystyrene nanospheres. Food Chemistry. 2019;**275**:377-384. DOI: 10.1016/j.foodchem.2018.09.111

[6] Zhou WH, Liu FM, Yi SP, Chen YZ, Geng XY, Zheng CM. Simultaneous stabilization of Pb and improvement of soil strength using nZVI. Science of the Total Environment. 2019;**651**:877-884. DOI: 10.1016/j.scitotenv.2018.09.146

[7] Shao JC, Yu XN, Zhou M, Cai XQ, Yu C. Nanoscale zero-valent iron decorated on bentonite/graphene oxide for copper ions removal from aqueous solution. Materials. 2018;**11**:945. DOI: 10.3390/ma11060945

[8] Yu XN, Qian CX, He ZH, Sun LZ. Removal of heavy metal Ba (II) by phosphate-mineralization microbe under different standing time. Fresenius Environmental Bulletin. 2017;**26**:859-863

[9] Qian CX, Zhan QW. Bioremediation of heavy metal ions by phosphate-mineralization bacteria and its mechanism. Journal of the Chinese Chemical Society. 2016;**63**:635-639. DOI: 10.1002/jccs.201600002

[10] Qian CX, Xiaoniu Yu XN, Wang X. Potential uses and cementing mechanism of bio-carbonate cement and bio-phosphate cement. AIP Advances. 2018;**8**:095224. DOI: 10.1063/1.5040730

[11] Li L, Qian CX, Cheng L, Wang RX. A laboratory investigation of microbe-inducing $CdCO_3$ precipitate treatment in Cd^{2+} contaminated soil. Journal of Soils and Sediments. 2010;**10**:248-254. DOI: 10.1007/s11368-009-0089-6

[12] Yu XN, Qian CX, Wang X. Biosynthesis of magnesium phosphates and its thermal property. Science of Advanced Materials. 2015;**7**:1730-1733. DOI: 10.1166/sam.2015.2392

[13] Wang X, Qian CX, Yu XN. Synthesis of nano-hydroxyapatite via microbial method and its characterization. Applied Biochemistry and Biotechnology. 2014;**173**:1003-1010. DOI: 10.1007/s12010-014-0911-5

[14] Yu XN, Qian CX, Wang X. Bio-inspired synthesis of barium phosphates nanoparticles and its characterization. Digest Journal of Nanomaterials and Biostructures. 2015;**10**:199-205

[15] Yu XN, Qian CX. Biological reduction-deposition and luminescent properties of nanostructured $CePO_4$@

$NaCe(SO_4)_2(H_2O)$ and $CePO_4$. Materials Chemistry and Physics. 2016;**171**:346-351. DOI: 10.1016/j.matchemphys.2016.01.026

[16] Yu X, Wang X, Qian C. Microbiological mineralization: Bacillus-induced Vivianite $Fe_3(PO_4)_2{\cdot}8H_2O$ precipitation. Digest Journal of Nanomaterials and Biostructures. 2014;**9**:1373-1378

[17] Nallamuthu D, Selvarajan P, Freeda TH. Studies on various properties of pure and Li-doped barium hydrogen phosphate (BHP) single crystals. Physica B: Condensed Matter. 2010;**405**:4908-4913. DOI: 10.1016/j.physb.2010.09.030

[18] Arora SK, Trivedi TR, Oza AT, Patel VA. IR spectroscopy of barium hydrogen phosphate. Acta Materialia. 2001;**49**:2103-2107. DOI: 10.1016/s1359-6454(01)00024-6

[19] Höppe HA, Daub M, Oeckler O. Synthesis, crystal structure, infrared spectrum and thermal behaviourof αlpha-BaHPO4. Solid State Sciences. 2009;**11**:1484-1488. DOI: 10.1016/j.solidstatesciences.2009.05.007

[20] Hebbar KC, Dharmaprakash SM, Rao PM. Growth of barium hydrogen phosphate crystals. Journal of Materials Science Letters. 1991;**10**:1430-1432. DOI: 10.1007/BF00724398

[21] Dinamani M, Vishnu Kamath P. Electrochemical synthesis of metal phosphates by cathodic reduction. Materials Research Bulletin. 2001;**36**:2043-2050. DOI: 10.1016/S0025-5408(01)00682-1

[22] Wang F, Xu GY, Zhang ZQ. The effect of pH on morphological control of barium hydrogen phosphate crystal by a double-hydrophilic copolymer. Materials Letters. 2005;**59**:808-812. DOI: 10.1016/j.matlet.2004.11.026

[23] Ji Y, Gao H, Sun J, Cai F. Experimental probation on the binding kinetics and thermodynamics of Au(III) onto Bacillus subtilis. Chemical Engineering Journal. 2011;**172**:122-128. DOI: 10.1016/j.cej.2011.05.077

[24] Yang M, You H, Zheng Y, Liu K, Jia G, Song Y, et al. Hydrothermal synthesis and luminescent properties of novel ordered sphere $CePO_4$ hierarchical architectures. Inorganic Chemistry. 2009;**48**:11559-11565. DOI: 10.1021/ic901829v

[25] Frost RL, Weier ML. Raman spectroscopic study of vivianites of different origins. Neues Jahrbuch fuer Mineralogie-Monatshefte. 2004;(10):445-463. DOI: 10.1127/0028-3649/2004/2004-0445

[26] Walpersdorf E, Koch CB, Heiberg L, O'Connell DW, Kjaergaard C, Hansen HCB. Does vivianite control phosphate solubility in anoxic meadow soils? Geoderma. 2013;**193-194**:189-199. DOI: 10.1016/j.geoderma.2012.10.003

Section 5

Applications of Nanomaterials

Chapter 8

SiO_2-Based Materials for Immobilization of Enzymes

Crina Anastasescu, Mihai Anastasescu, Ioan Balint and Maria Zaharescu

Abstract

It is well known that SiO_2-based nanomaterials were widely used as support material for many chemically active species but also for compounds with biological activity such as antibodies and enzymes, due to their lack of toxicity and high surface area. The hybrid materials resulted from associating enzymes with various morphologies of SiO_2 inorganic matrix, especially obtained by sol-gel method, are meant to develop a higher enzymatic activity and increased lifetime but also the recovery and reutilization advantage. The present contribution emphasized the synthesis of SiO_2 nanomaterials with different morphologies and their physicochemical characteristics including bio-catalytic activity of immobilized enzymes on simple SiO_2. The morphology-dependent behavior of SiO_2 inorganic carriers obtained by sol-gel method was also emphasized. Accordingly, SEM investigations, nitrogen sorption, electrokinetic potential, and spectroscopic measurements are presented. p-Nitrophenyl acetate (p-NPA) was used for testing the enzymatic activity of as-prepared lipase-SiO_2 hybrid materials.

Keywords: SiO_2, sol-gel, enzymes, immobilization

1. Introduction

New multifunctional materials are required as practical solutions for competitive, affordable, and environmentally friendly technologies from industrial and health-care areas. These should be also able to sustain and preserve various ecosystems and an appropriate quality of human life, gathering resources from border scientific fields: materials science, bioremediation, biocatalysis, and biomedicine [1].

Silica is largely studied material with high chemical and thermal stability, nontoxic and non-expensive, which can display a huge variety of morphologies, not only in natural state but also obtained by various synthetic routes. For these reasons, silica with high surface area (MCM 41, SBA 15) is intensively used as support for chemical active species [2] but also as inorganic host in hybrid complex for biological active compounds (enzymes) [3], especially those obtained in mild conditions, by sol-gel method [4]. These materials are recognized as appropriate carriers for biological compounds, preserving or increasing their intrinsic features [5]. In addition, tubular morphology of silica allows custom modification/functionalization of internal/external surfaces with chemical (metals, oxides) and biological (enzymes, antibodies) active compounds [6].

In this work we review various SiO_2 morphologies obtained during our material development researches, pointing out their peculiar properties and advantages related to the immobilization of enzymes for potential practical applications.

2. Sol-gel synthesis of different SiO_2 structures obtained by sol-gel method: sphere-, tube-, and veil-type morphologies

Our previous works reported sol-gel synthesis and characterization of SiO_2 matrices with different morphologies (nanotubes, spheres, and veils) but also emphasized the multifunctional behavior of the as-obtained SiO_2 materials, from the perspective of their chemical and biological activity. Firstly, spherical particles and hollow nanotubes with open ends, square shape, and high surface area (300 m^2/g), [7, 8] were obtained at room temperature and atmospheric pressure, using DL tartaric acid as in situ generated template. The catalytic [9] and photocatalytic [10] activities of platinum-modified SiO_2 nanotubes and spheres were further investigated and proven to be dependent on morphology of SiO_2 matrix which bears (supports) the metallic active phase. SiO_2 veils have been also obtained, and their composites have been used as support for platinum electrocatalyst [11].

Improvement of in situ generated template-assisted SiO_2 synthesis and appropriate post-synthesis treatments allowed us to obtain nanotubes with intrinsic photo- and electrochemical activity due to high density of oxygen-related lattice defects [12, 13] but also to reveal the distinctive capacity of highly defected SiO_2 and its composite SiO_2-TiO_2 to generate reactive oxygen species (ROS) with potential applications in biology and medicine [14].

SiO_2 materials (nanotubes) have been proven to be biologically active for bioremediation processes, showing bactericidal and bacteriostatic effects on halotolerant microorganisms [15]. These type of materials have been successfully used as inorganic carriers in hybrid complex with enzymes [16], the enzymatic activity of the complex being dependent on morphological and textural properties of SiO_2 matrices [17].

Generally, sol-gel synthesis of tubular SiO_2 materials was conducted in line with Nakamura and Matsui's work [18], modifications of synthetic procedure being done in respect to particular applications.

Thin and larger SiO_2 nanotubes in addition to spherical SiO_2 are meant to function as inorganic host for immobilized enzymes, and, accordingly, their previously synthesis has been reported [7, 17]. Better control of synthesis conditions allowed an optimization of their morphology and structure, in order to be appropriate for an inorganic carrier. Generally, these materials are resulting from a template-assisted sol-gel process, consisting in hydrolysis and condensation reactions conducted in the presence of DL tartaric acid, in respect to well-defined and well-controlled synthetic parameters which will be exposed in what comes. By selecting proper temperature, concentration of started solution, and in situ organic template development, not only morphological changes (observed in SEM captured images) can be induced but also structural ones, resulting in highly defected surfaces with increased chemical reactivity [12].

In this sense, we used a mixture of tetraethyl orthosilicate (TEOS) (Alfa Aesar, 99%), absolute ethanol (Riedel-de Haen), ammonia (25% w/w aq. Alfa Aesar), and DL tartaric acid (TA) as internal templating agent. The involved molar ratio was 1 (TEOS):0.07 (tartaric acid):43 (ethanol):36 (H_2O), at 20°C. A subsequent addition of ammonia (25%), drop by drop, triggers the formation of the final synthesis product, a white and dense suspension being observed. The whole synthetic procedure is performed in mild conditions, at room temperature (25°). One hour aging time is allowed followed by filtering, washing with ultrapure water (Milli-Q system, 18 MΩcm), and drying at 100°C for 3 hours. Typically, the post-synthesis thermal treatment involves calcination for 3 h at 500°C with a heating rate of 0.5°C/minute (**Figure 1**).

This synthesis route leads to hollow and large nanotubes, together with a minor percent of spherical particles. The sample will be denoted in what comes as SiO_2TL sample (TL meaning large tubes) (**Figure 3**).

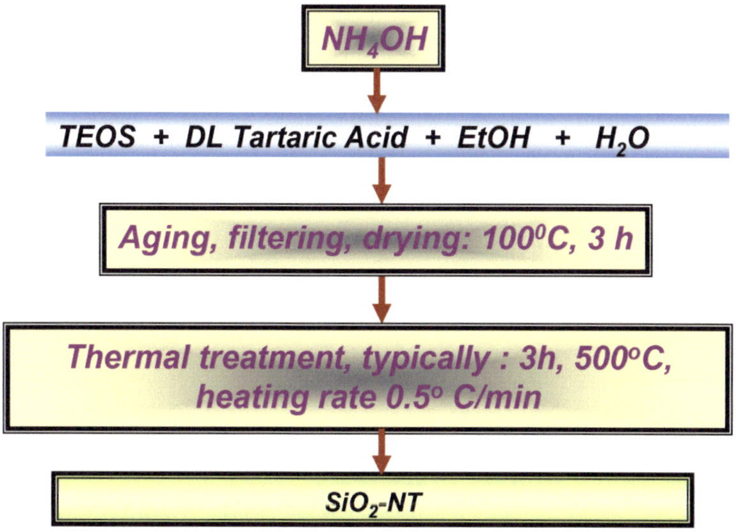

Figure 1.
Schematic representation of the SiO_2-NT preparation route.

A faster addition of aqueous ammonia to similar reaction mixture favors the formation of solid spheres instead of nanotubes. Their diameters are about hundreds of nanometers, the samples being denoted as SiO_2S (S—spheres) (**Figure 4**).

For obtaining thinner (20–50 nm diameter) and longer nanotubes, denoted as SiO_2Tt (Tt meaning thin tubes) (**Figure 5**), the working temperature decreased at 0°C, and the molar ratio of reactants changed according to the following values: 1 (TEOS):0.03 (tartaric acid):18 (ethanol):10 (H_2O).

These synthetic conditions allow to obtain not only a better morphological homogeneity than in previous case but also an increased chemical reactivity of the SiO_2 tubular surfaces; their functionalization with NH_2 groups proved to be favored, which is an appropriate functional feature for an efficient inorganic carrier.

The control of the above-described synthesis parameters allows to achieve various and reproducible morphologies. Briefly, the diversity regarding the morphology (spheres and veils) but also the different textural properties of the nanotubes is closely related to the experimental conditions governing both the SiO_2 sol and in situ template generation.

Accordingly, adjustments concerning the SiO_2 sol formation are resulting in appearance of the veil-type morphology (**Figure 6**).

In addition, by using *meso*-tartaric acid instead of DL tartaric acid, small spherical particles (around 40 nm) have been obtained (**Figure 7**).

If DL tartaric acid is replaced by oxalic acid, tubular morphology is maintained. **Figure 8** shows the resulted tubes sized (characterized) at micrometric scale both in diameter and length. These experimental achievements are in line with the literature data [19].

3. Modification of SiO_2 tubular matrices with gold nanoparticles

According to literature data, most of the enzymatic immobilization tests on silica are based on electrostatic interactions, resulting in direct adsorption [20, 21]. Gold nanoparticles deposed on silica surface proved to increase the capacity of SiO_2 materials to load the target enzymes [22].

In this sense, gold modification of SiO_2 matrices was performed both by direct impregnation of SiO_2 powder with a solution of $HAuCl_4$ $3H_2O$ and by

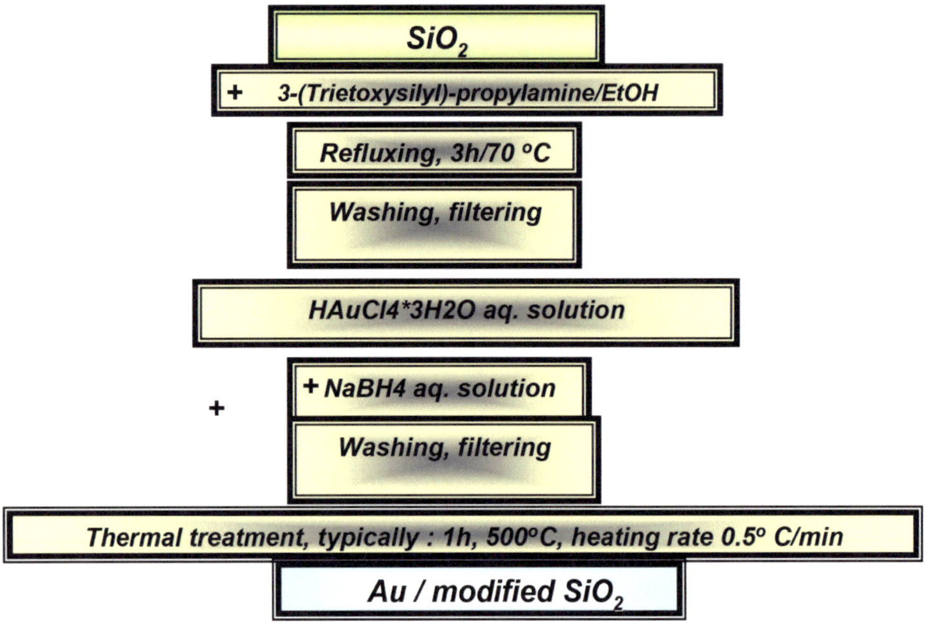

Figure 2.
Schematic representation of the Au-modified SiO_2-NT preparation route.

deposition of metallic nanoparticles after functionalization of silica surface with 3-(triethoxysilyl)-propylamine.

Direct modification of SiO_2 materials with gold was done using 0.2 g of previously thermally treated SiO_2 (500°C, 3 hours) and an aqueous solution of 0.01 M $HAuCl_4$ $3H_2O$ (49%, Sigma-Aldrich), in order to achieve 2% Au (weight percent). The resulted mixture has been stirred (500 rpm) at room temperature for 5 hours and then calcined 1 h at 300°C with a heating rate of 0.5 deg/min.

The same procedure has been respected for gold modification of SiO_2 surface after functionalization with organic groups (**Figure 2**), which consists in refluxing for 12 hours of a mixture containing 0.2 g SiO_2, 5 ml absolute ethanol, and 50 microliters (μl) of 3-(triethoxysilyl)-propylamine ($C_9H_{23}NO_3Si$, 99%, Sigma-Aldrich).

After washing with ethanol, filtering, and drying, the recovered powder is added to a gold aqueous solution ($HAuCl_4$ $3H_2O$, 0.01 M) and stirred (500 rpm) at room temperature for 5 hours. An aqueous solution of $NaBH_4$ (Sigma-Aldrich) is used for reduction of ionic (Au^{3+}) to metallic gold, followed by washing, filtering, and calcinations of the recovered product at 500°C for 1 hour, maintaining the same heating rate as for direct impregnation (0.5 deg/min).

The average size of deposited metal nanoparticles evaluated from TEM and X-ray diffraction (XRD) measurements ranged in 5–15 nm domain.

4. Morphological and structural characterization of simple and gold-modified SiO_2

In order to evaluate the various morphologies displayed by SiO_2 sol-gel materials as a result of ranging the synthetic parameters and to emphasize the fully desired characteristics of SiO_2 matrices for development of hybrid structures with enzymatic activity, scanning and transmission electron microscopies (SEM, TEM), Fourier transform infrared and FTIR-ATR, diffuse reflectance UV-Vis spectroscopy, and XRD measurements were performed.

4.1 Characterization of bare SiO_2 materials

4.1.1 Scanning and transmission electron microscopy (SEM and TEM)

SEM investigations (**Figure 3**, for the SiO_2TL sample) clearly show the tubular morphology as major phase of the SiO_2 powders resulted from the sol-gel process.

SEM images from **Figure 3a–d** reveal a small percent of rounded particles, most of them being attached on the external surface of the tubes. This observation is in line with reported literature data about SiO_2 tubular matrices obtained in mild conditions, by sol-gel process [18]. Their average external diameter is around 200 nm, the length overcoming 1 μm. An accurate perspective on the wall structure of these large nanotubes (SiO_2TL) is achieved by TEM images from **Figure 3e, f**. So, thick walls with porous structure, empty core, and open ends of the nanotubes could be observed.

Figure 4 shows the presence of solid, big spheres with an average diameter of 200 nm, no other morphologies being present.

Figure 5 presents thin nanotubes (SiO_2Tt) (average diameter ~50 nm) like predominant morphology, spherical entities being present on their surface but as smaller percent than for the previous tubular sample (SiO_2TL). From this point of view, it is obvious that the sample has a high degree of homogeneity. Also, the tubes

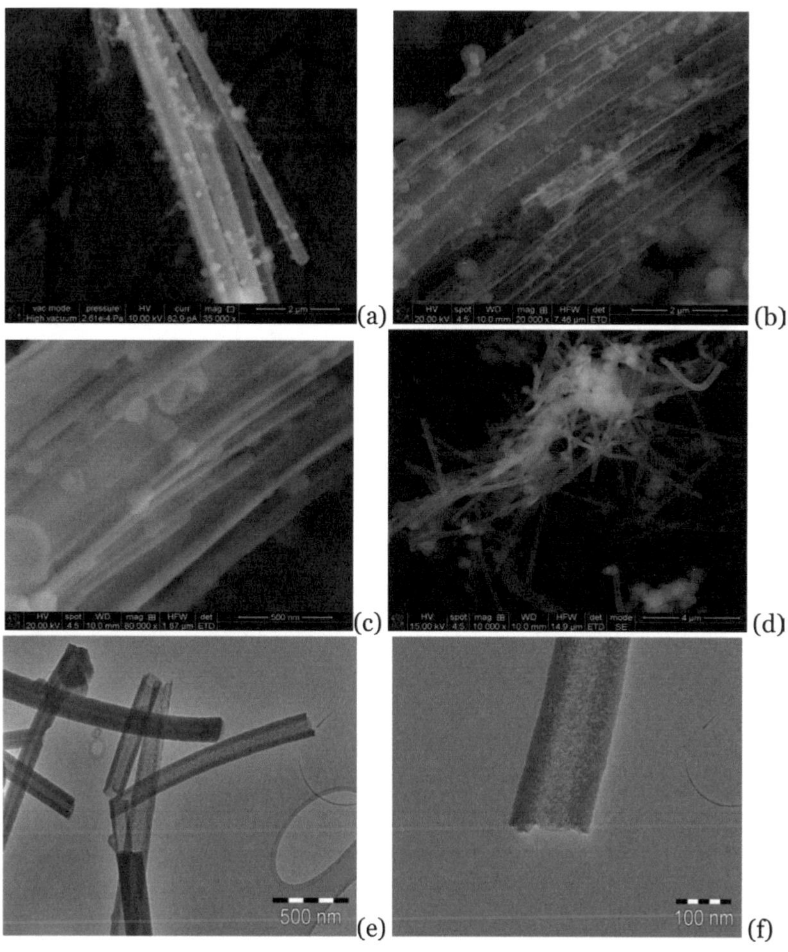

Figure 3.
(a–d) SEM images at different magnifications of large tubes (SiO_2TL)—unpublished results; (e, f) TEM images of large tubes (SiO_2TL). (e, f) (Reprinted from [7] with permission from Elsevier).

are thinner and much longer than previous ones, the ratio between the external and internal diameters being also decreased. Despite their impressive length, similar to that of the fibers, the SEM images captured at 1 μm scale bar show a good contrast between their walls and empty cores. This means high surface (both internal and external) areas are available for interaction (contact) with enzymes. In addition to BET results, these images indicate a facile access but also a potential confinement [23] of the enzymes inside the tubes which could favor an increased degree of enzymatic immobilization.

Figure 6 presents images of a single and distinct morphology obtained by sol-gel method, namely, SiO_2 veils. In addition to the previously reported application, as support and part in composite structures [11], many potential applications could exploit the lack of toxicity but especially the great structural and morphological homogeneity of this SiO_2 matrix, especially in biomaterial range.

Figures 7 and **8** are also relevant for the influence of the synthesis parameters on the morphological features of the synthesized material. The change of the organic template leads to rounded, quite undefined nanoparticles characterized by the SEM images from **Figure** 7 and the modified tubular morphology being exposed in **Figure** 8. These images are recorded for a sample whose characteristics have been tailored by the oxalic acid used as template. Huge tubes can be observed and enlarged especially as diameters (up to 5 μm).

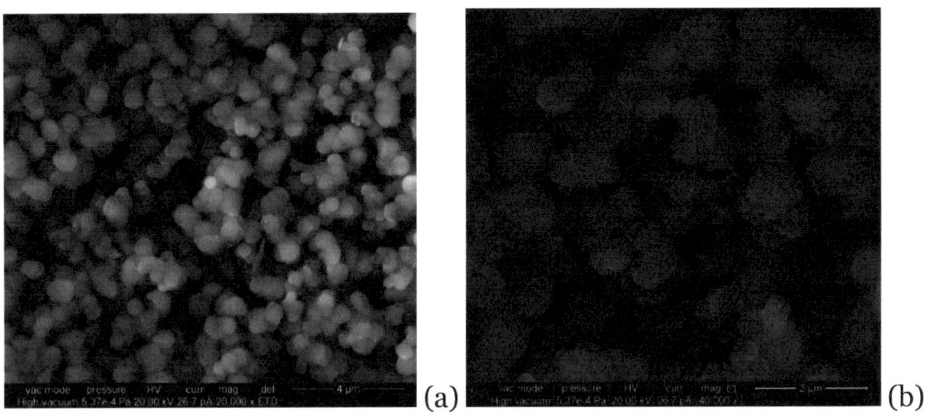

Figure 4.
SEM images of SiO_2 spheres (SiO_2S). (Reprinted from [17]).

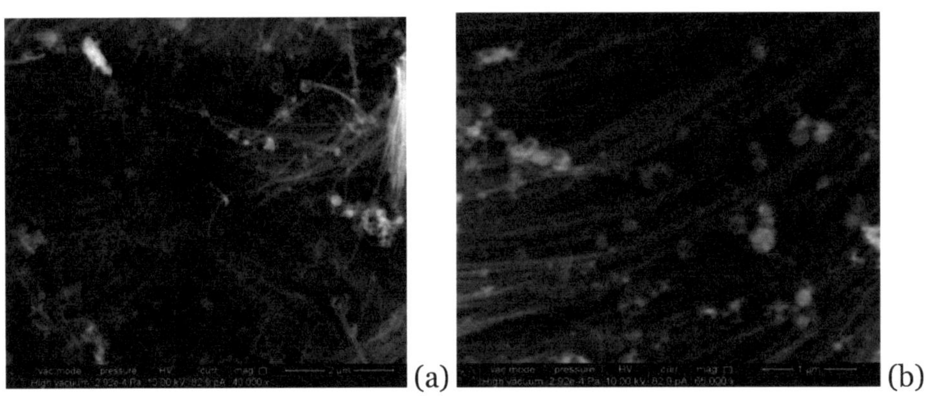

Figure 5.
SEM images of thin tubes (SiO_2Tt). (Reprinted from [17]).

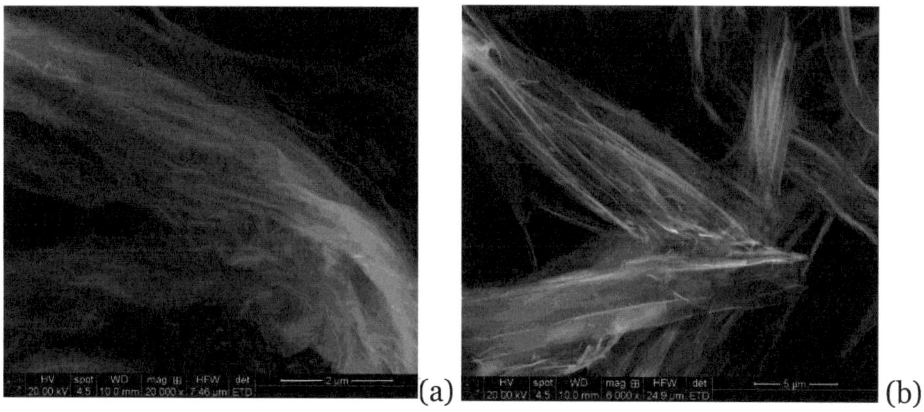

Figure 6.
SEM images of sol-gel SiO_2 veils (SiO_2v). (Unpublished results).

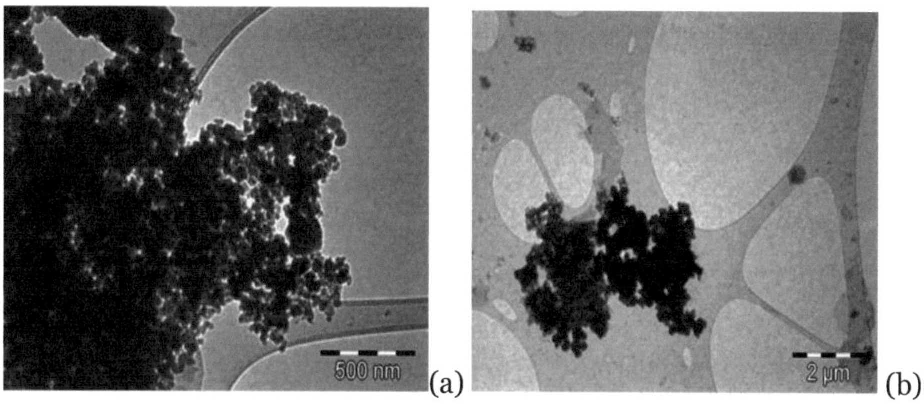

Figure 7.
TEM images of spherical SiO_2 nanoparticles (SiO_2s). (Reprinted from [7] with permission from Elsevier).

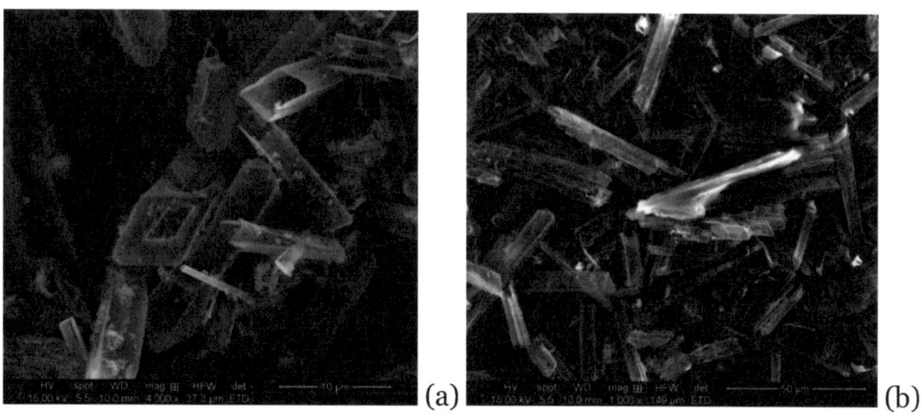

Figure 8.
SEM images of extra large SiO_2 tubes (SiO_2TXL) obtained with oxalic acid as template. (Unpublished results).

4.1.2 Fourier transform infrared spectroscopy (FTIR)

Taking into account the target application for these SiO_2-based materials, namely, the immobilization of enzymes, the information obtained by FTIR regarding the structural characteristics of SiO_2 inorganic matrix and its interaction with functional organic groups are useful.

The relevant vibration bands observed from FTIR spectra of the spherical and tubular SiO_2 samples (**Figure 9**) are similar; small variations could be perceived especially regarding their intensity and the presence of silanol groups (Si-OH), which is an important issue from the perspective of further functionalization of silica surface with 3-(triethoxysilyl)-propylamine (APTES) and, subsequently, the deposition of gold on the surface.

According to our previous collected results and literature data, the large peak from 1124 cm^{-1} is due to Si-O stretching from SiO_2 [7], and its shoulder (1200 cm^{-1}) is assigned to asymmetric vibration of Si-O-Si [24]. One more tiny shoulder located around ~ 960 cm^{-1} could indicate the presence of silanol groups [25], especially for spherical SiO_2 (SiO_2S) but unperceivable for the sample containing large tubes (SiO_2TL). Its spectrum indicates also less structural, free OH groups and adsorbed water, generally evaluated from the development of a broadband ranging in 3350–3600 cm^{-1} domain [26, 27]. The presence of OH groups on silica surface could promote the hydrogen bonding formation with functional groups of enzymes [28]. According to the reported data [8], the peak located at 847 cm^{-1} is due to $\nu_{s(Si\text{-}O\text{-}Si)}$, and the peak appearing to 484 cm^{-1} is determined by $\delta_{(Si\text{-}O\text{-}Si)}$.

In order to identify and characterize gold nanoparticles deposited on the surface of thin tubes by direct impregnation of powder with gold precursor solution, SEM analysis (**Figure 10a**) was performed. A very large size distribution associated with a weak dispersion of gold nanoparticles could be observed.

In the case of gold deposition after surface functionalization with APTES, smaller gold nanoparticles were expected to result. **Figure 11a** presents a SEM image of the metal-modified nanotubes, but gold nanoparticles are unperceivable because of their reduced dimensions as a result of previous surface functionalization. Accordingly, TEM analysis was performed, and the images from **Figure 11b** show smaller gold nanoparticles with average diameter around 7 nm, deposited on external surface of nanotubes which bear as well spherical particles of SiO_2. Also, a loss of transparency could be observed in the case of functionalized samples. Images from **Figures 10** and **11** emphasize more efficient gold deposition after functionalization of SiO_2 tubular surface, exhibiting small, well-dispersed gold nanoparticles which should successfully bind the enzymes.

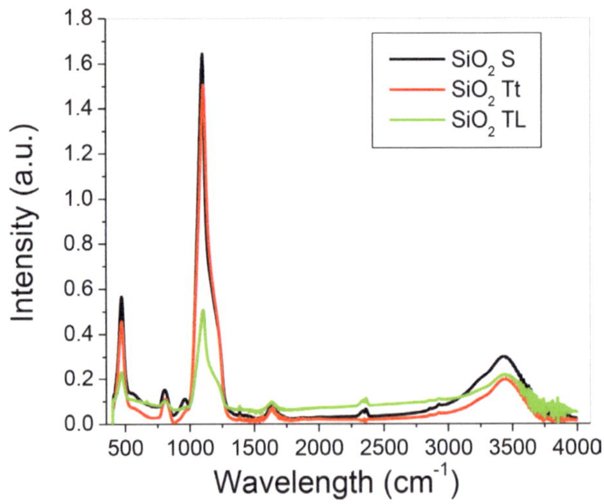

Figure 9.
FTIR spectra of SiO_2 spheres (SiO_2S), SiO_2 thin tubes (SiO_2Tt), and SiO_2 large tubes (SiO_2TL).

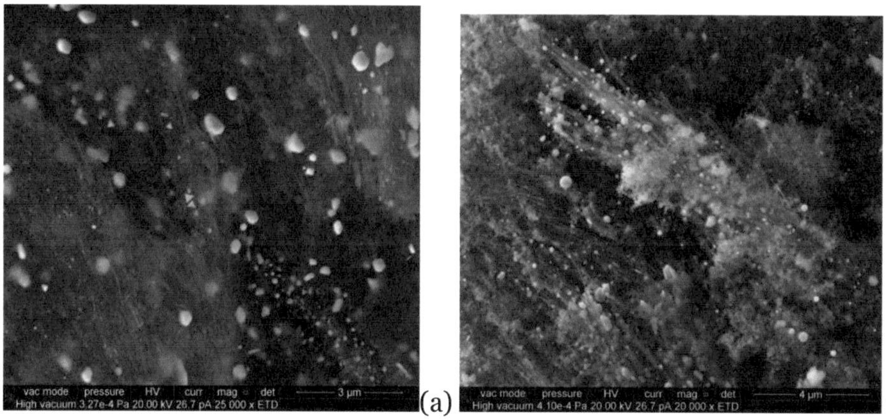

Figure 10.
SEM images of gold-modified silica thin tubes, by direct impregnation ($AuSiO_2Tt$). (Unpublished results).

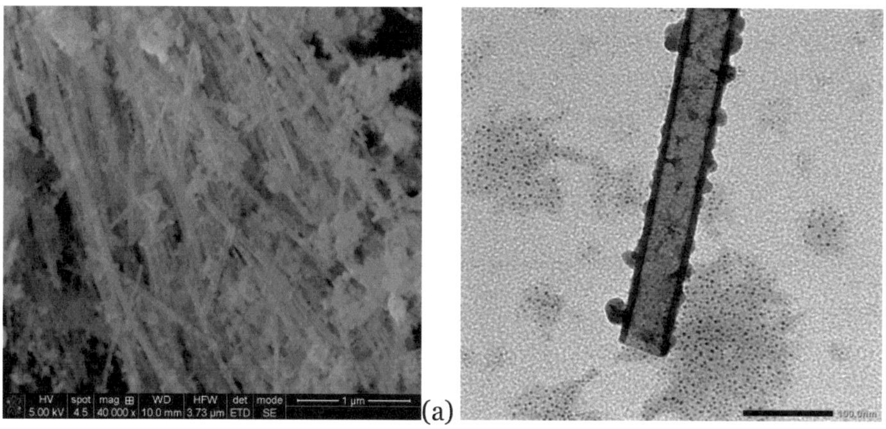

Figure 11.
(a) SEM images of SiO_2 thin tubes modified with gold after functionalization with APTES ($AuSiO_2TtF$). (b) TEM images of SiO_2 thin tubes modified with gold after functionalization with APTES ($AuSiO_2TtF$). (Unpublished results).

4.1.3 Diffuse reflectance UV-Vis spectroscopy performed on silica and gold-modified silica thin tubes

The main objective of this investigation was to observe the surface plasmon resonance (SPR) phenomenon which is related to gold nanoparticles contained by the analyzed samples. As it could be seen from **Figure 12**, a large peak appears around 500 nm for the gold-modified SiO_2 samples, this being more impressive when gold is deposited after functionalization of SiO_2 thin tubes ($AuSiO_2TtF$).

4.1.4 X-ray diffraction (XRD)

The XRD results are presented in **Figure 13**, where the broad diffraction band located at $2\Theta = 21.5^{\circ}$ is characteristic to the amorphous structure of SiO_2 and the maxima from $2\Theta = 38.17, 44.3, 64.5, 77.5^{\circ}$ are typical for metallic gold. The average crystallite size of the gold deposited after functionalization of silica surface, calculated with Williamson-Hall method, is 7.5 nm, much smaller than for the other sample (~ 15 nm).

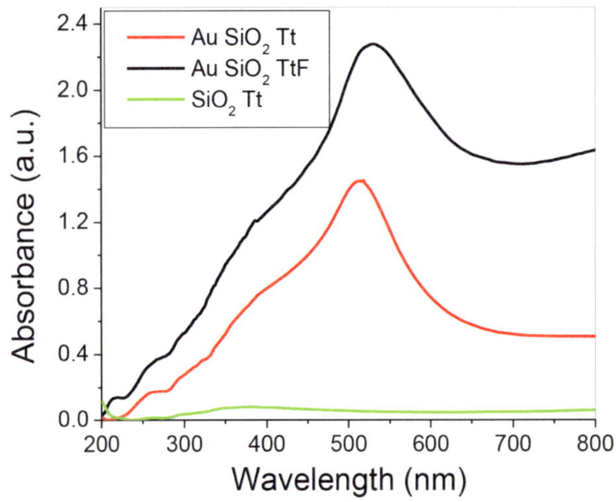

Figure 12.
UV-Vis spectra of silica and gold-modified silica tubes by direct impregnation ($AuSiO_2Tt$) and after functionalization ($AuSiO_2TtF$). (Unpublished results).

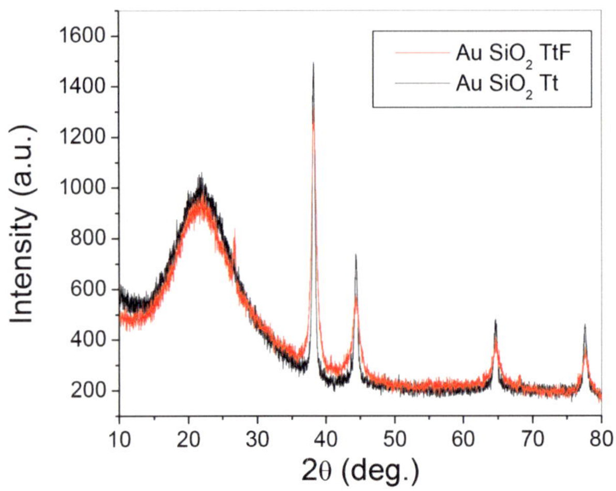

Figure 13.
XRD spectra of gold-modified silica thin tubes by direct impregnation ($AuSiO_2Tt$) and after functionalization ($AuSiO_2TtF$). (Unpublished results).

4.1.5 Fourier transform infrared spectroscopy (FTIR)

Analysis of spectra from **Figures 15** and **16** (details of general spectrum from **Figure 14**) shows that after silica thin tube functionalization, its spectra do not present the specific vibrations of Si-OH (~ 960 cm^{-1}), free or bounded OH groups, and adsorbed water (3350–3600 cm^{-1}). This supports the idea that 3-(triethoxysilyl)-propylamine is anchored by the OH groups of the tubular silica surface.

4.1.6 FTIR-ATR spectroscopy performed on functionalized (APTES) SiO_2 thin tubes

FTIR-ATR measurements were done in order to check the functionalization of SiO_2Tt surface with functional groups of 3-(triethoxysilyl)-propylamine.

Figure 17 identifies the peak located at 1635 cm^{-1} which, according to the literature data, is attributed to the twisting vibration of N-H in addition to the band from 3000 to 2900 cm^{-1} due to the stretching vibration of the CH_2 which proves the successful functionalization of silica with APTES [29, 30].

It is obvious from **Figure 18** that the chemical "fingerprint" of the SiO_2 inorganic matrix is preserved, but the organic groups identified in previous spectrum are strongly flattened, due to the thermal treatment applied after functionalization and gold modification, respectively. As expected, the broadband (3400–3600 cm^{-1}) related to OH groups and adsorbed water is clearly diminished, as well as the band from 2900 to 3000 cm^{-1} (the stretching vibration due to CH_2 from APTES).

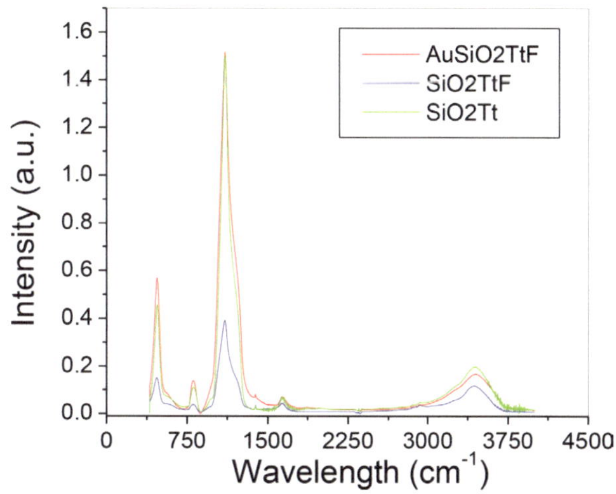

Figure 14.
General FTIR spectra of simple, functionalized, and gold-modified SiO_2 thin tubes: SiO_2 (SiO_2Tt), SiO_2-APTES (SiO_2TtF), and $AuSiO_2$-APTES ($AuSiO_2TtF$). (Unpublished results).

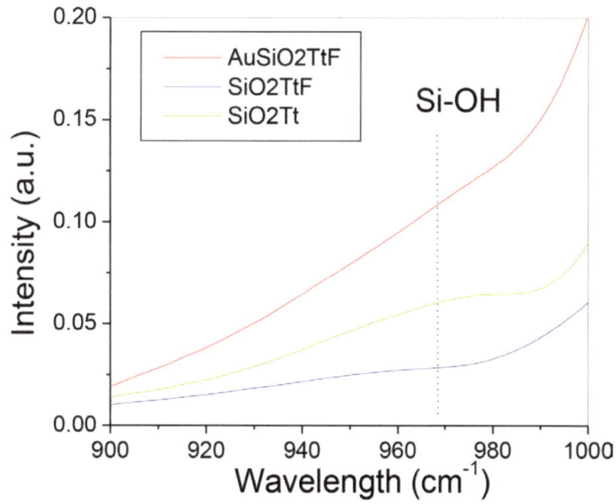

Figure 15.
FTIR spectra of simple, functionalized, and gold-modified SiO_2 thin tubes: SiO_2 (SiO_2Tt), SiO_2-APTES (SiO_2TtF), and $AuSiO_2$-APTES ($AuSiO_2TtF$) in 900–1000 cm^{-1}region. (Unpublished results).

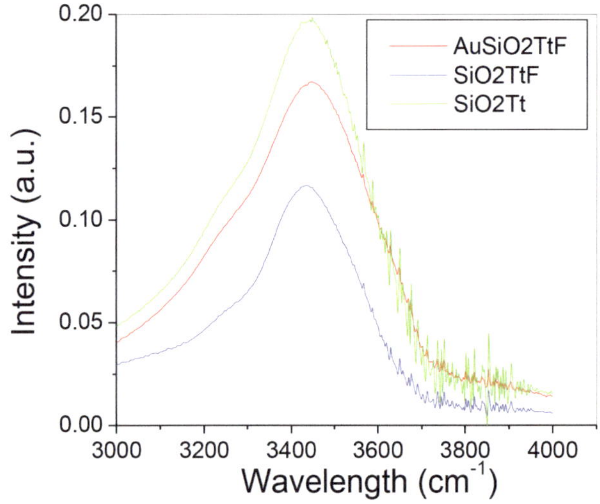

Figure 16.
FTIR spectra of simple, functionalized, and gold-modified SiO_2 thin tubes: SiO_2 (SiO_2Tt), SiO_2-APTES (SiO_2TtF), and $AuSiO_2$-APTES ($AuSiO_2TtF$) in 3000–4000 cm^{-1}region. (Unpublished results).

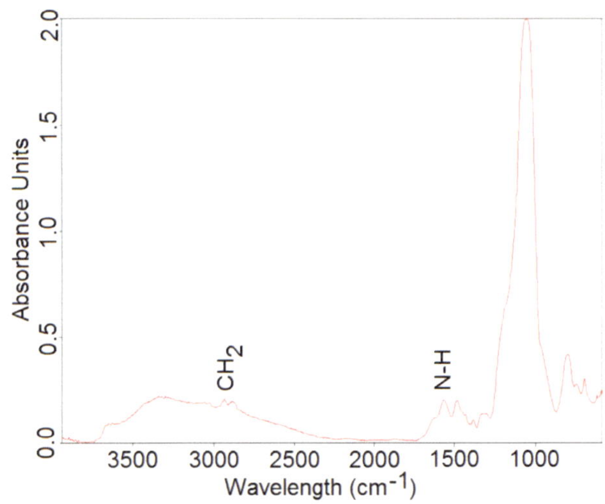

Figure 17.
FTIR-ATR spectrum of thin tubes (SiO_2TtF) functionalized with 3-(triethoxysilyl)-propylamine. (Unpublished results).

5. Synthesis, characterization, and catalytic activity of hybrid complex lipase-inorganic matrices (SiO_2, $AuSiO_2Tt$)

5.1 Synthesis of lipase-SiO_2 complex

Generally, the design of hybrid structures like enzyme-inorganic carrier is looking to preserve or increase the catalytic activity of free enzymes but especially to extend its lifetime and endurance in relation to parameters governing its proximity (pH, temperature) [31]. In addition to this, enzymatic immobilizations have the advantage of reusability which is an important issue especially for those biocatalysts meant to sustain valuable biotechnologies. For instance, free and immobilized

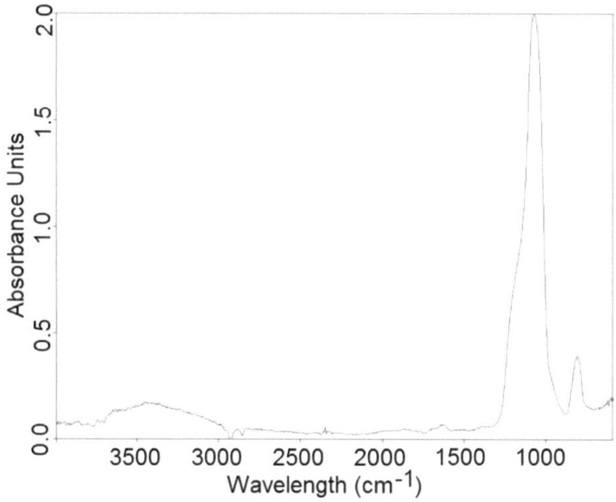

Figure 18.
FTIR-ATR spectrum of gold-modified thin tubes ($AuSiO_2TtF$) after functionalization with 3-(triethoxysilyl)-propylamine. (Unpublished results).

lipase is used in esterification, transesterification, reactions [32, 33], and different biotechnological applications [34].

According to literature data, gold nanoparticles could increase the enzyme loading on inorganic carriers, acting as efficient support for enzymes [35]. Few but important potential medical applications are centered on the use of lipase immobilized on gold nanoparticles as electrochemical sensor for triglycerides from human serum [36].

The literature results indicate the isoelectric point of lipase from *Rhizopus oryzae* and silica at pH 7.6 and 3, respectively [21]. In order to achieve an electrostatic attraction between SiO_2 supports and lipase, our immobilization experiments were conducted at pH 6.3 where lipase is slightly positively charged.

Briefly, hybrid complex material consisting in lipase-inorganic support (SiO_2Tt, $AuSiO_2Tt$) was obtained according to the previous reported work [17] concerning the generation of biocatalyst lipase-SiO_2 matrices. Lipase (*Rhizopus oryzae*, Sigma-Aldrich) in phosphate buffer (pH 6.3) was mixed with SiO_2 and Au/SiO_2 powder (0.02 g) also dispersed in phosphate buffer and magnetically stirred 24 h (100°C). The solid product containing lipase immobilized on SiO_2 and Au/SiO_2 powder was separated from supernatant by centrifugation and washed several times with phosphate buffer and water in order to be further evaluated from the perspective of total organic carbon content, FTIR-ATR and PZC measurements, and enzymatic tests as well.

5.2 Characterization of lipase-SiO_2 complex

5.2.1 BET measurements

The textural properties of tubular and spherical silica, correlated with total organic carbon measurement on their lipase derivative hybrid complex, were previously evaluated and reported [17].

Accordingly, a surface area (S_{BET}) about 18 and 14m^2/g, respectively, was found for large tubes (SiO_2TL) and spherical (SiO_2S) samples. In the case of large tubes, mesopores of 3 nm were identified, which are probably located in the walls of the nanotubes but improper for the lipase immobilization (the diameter of *Rhizopus oryzae* being ~ 5 nm). SEM and TEM images (**Figure 3**) show an internal diameter of ~200 nm, and total organic carbon measurements [17] indicate the best lipase loading on SiO_2Tt sample (30 mg/g). It is obvious that the internal surfaces are accessible for lipase immobilization.

Spherical SiO_2 matrices (SiO_2S) show a large size distribution of pores, ranging from 3 to 5 nm and from 10 to 30 nm. Probably, the last value is due to voids of packed spheres and could hardly allow the enzymatic immobilization, but leaching could be favored too.

Figure 19 presents the data related to SiO_2 tubular thin tube sample (SiO_2Tt).

From BET analysis, the value of the specific surface area (S_{BET}) was found to be 107.83 m^2/g. The shape of adsorption and desorption branches is typical for cylindrical pores with open ends. The pore size distribution is very large, spanning from 10 to 100 nm. By correlating these results with SEM images (**Figure 5**), it can be assumed that the internal diameters of these thin tubes are ranging between 10 and 50 nm. The well-defined peak from 3 nm can be attributed to the mesoporosity of wall nanotubes, similar to larger ones (SiO_2TL). These properties are drastically changed by the lipase adsorption; the broad peak ranging between 10 and 100 nm is not present anymore. This proves that the immobilization of lipase on SiO_2Tt sample is in fact a confinement inside the core of the nanotubes which can assure an increased stability against post-immobilization leaching. Total organic carbon analysis sets the capacity of loading the lipase around 15 mg/g, lower than for larger tubes but better than for spherical ones.

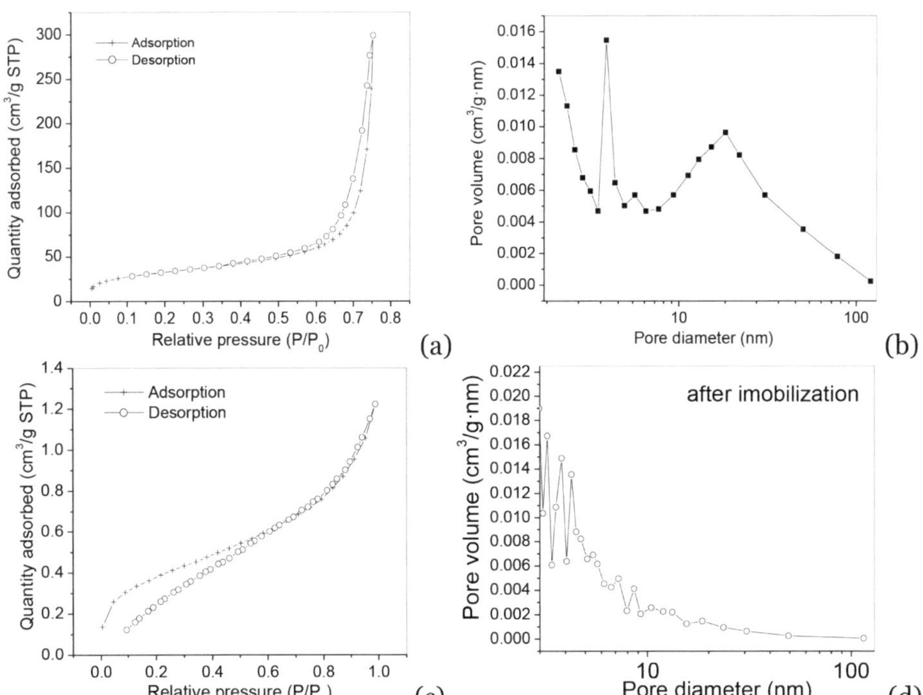

Figure 19.
N_2 adsorption-desorption isotherms (a) and the pore size distribution obtained from the desorption branch (b) for the SiO_2Tt samples before lipase immobilization; (c and d) the same curves recorded after lipase immobilization. (Reprinted from [17]).

5.2.2 FTIR-ATR measurements

Our reported data containing the FTIR-ATR results registered on lipase-SiO_2 hybrid complex [17] indicated the presence of lipase and its specific vibration bands both in supernatant and onto solid tubular and spherical SiO_2 supports (**Figure 20**).

The previous reported capacity of unmodified SiO_2 matrices to load lipase was found to fit to the following sequence: SiO_2TL > SiO_2Tt > SiO_2S (**Figure 20a**).

The comparison of **Figure 20a, b** highlights especially the peak from 3271 cm^{-1} (marking the stretching vibration of the NH group). In this sense, it appears that lipase from supernatant solution (of the derivative hybrid complex of gold-modified SiO_2Tt) is less emphasized. This observation means a higher capacity of gold-modified sample to uptake lipase than the bare SiO_2Tt carrier. Thus, the gold modification of silica looks to be successful and worth to be further investigated.

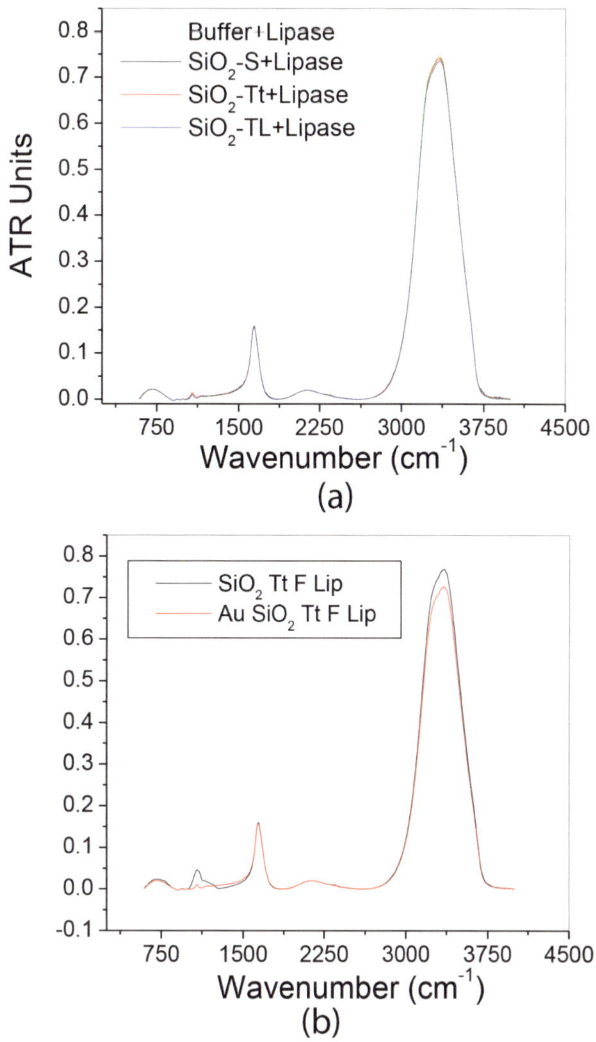

Figure 20.
(a) FTIR-ATR spectra recorded from supernatant solution after immobilization of lipase on different SiO_2 samples in phosphate buffer solution, revealing the 3271 cm^{-1} peak. (Reprinted from [17]); (b) FTIR-ATR spectra recorded from supernatant solution after immobilization of lipase on SiO_2Tt and AuSiO_2Tt supports. (Unpublished results).

5.2.3 The electrokinetic potential

The electrokinetic potential (ξ) measurements (**Figure 21**) bring information related to the electrostatic interactions established between the surfaces of inorganic carrier and enzyme, which is actually a decisive parameter for an efficient enzymatic immobilization by direct adsorption.

The lipase immobilization on SiO_2 and gold-modified SiO_2 matrices was performed at pH 6.3 in phosphate buffer solution. In these conditions, the silica should be negatively charged and lipase slightly positive. Our previous results [17] indicate highly negative values of electrokinetic potential (ξ), especially for spherical samples. After immobilization (**Figure 21a**), these values slightly change, probably due to the lipase adsorption on the external surface of SiO_2 matrices. From the above-presented data, the immobilization of lipase on SiO_2Tt sample takes place especially inside the nanotubes.

Figure 21b displays the values of electrokinetic potential (ξ) for gold-modified thin nanotubes, both by direct impregnation (AuSiO_2Tt) and after

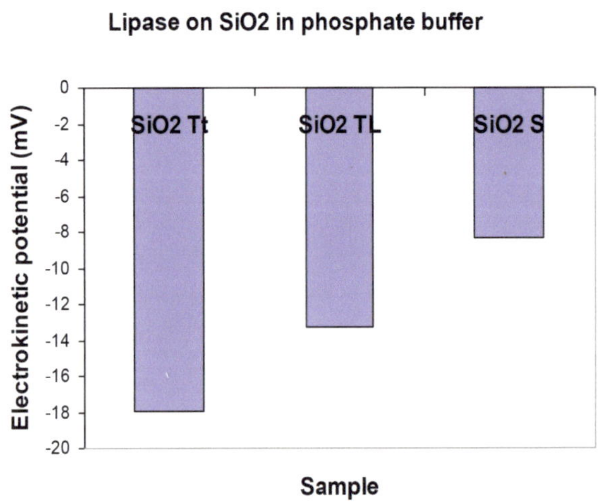

(a)

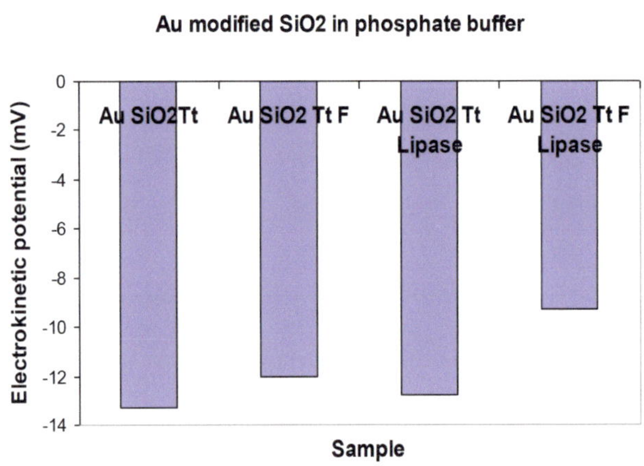

(b)

Figure 21.
(a) The electrokinetic potential of lipase-bare SiO_2 hybrid systems. (Adapted from [17]); (b) Gold-modified SiO_2Tt together with developed lipase hybrid complex. (Unpublished results).

functionalization with APTES ($AuSiO_2TtF$), before and after lipase immobilization. It is obvious that the deposition of gold after surface functionalization is more efficient in terms of the inorganic host ability to immobilize the lipase, this fact being illustrated in **Figure 21b** as a shift of the electrokinetic potential (ξ) to more positive values.

5.3 Catalytic activity of lipase-SiO_2 complex

Our previously reported data [17] are concerning the catalytic activity of lipase from *Rhizopus oryzae* immobilized on bare tubular and spherical SiO_2 matrices for hydrolysis reaction of p-nitrophenyl acetate (p-NPA).

Briefly, the hybrid complex newly formed is suspended in dimethyl sulfoxide (DMSO) and phosphate buffer at pH 7.8. One hour of incubation at 37°C is allowed, and the liquid phase was separated by centrifugation for measuring the reaction products by UV-Vis.

The best biocatalyst in terms of highest product amount (p-nitrophenyl) proved to result from lipase immobilization on SiO_2 thin tubes. For this reason, gold nanoparticle modification of this inorganic matrix should result in optimization of structural and functional properties of its derivative hybrid complex with lipase. This aspect was addressed by the above-presented structural characterization of gold-modified SiO_2Tt support, and its derivative complex and further investigations regarding their enzymatic activity will be performed.

6. Conclusions

Sol-gel method proves to be an efficient tool for obtaining well-defined SiO_2-based multifunctional materials (bio-/photocatalysis, bioremediation processes, electrochemistry).

SiO_2 inorganic carriers obtained by in situ generated template-assisted sol-gel method revealed morphology-dependent behavior regarding lipase (*Rhizopus oryzae)* immobilization and enzymatic activity of the derivative hybrid complex.

Gold modification of functionalized (APTES) SiO_2 tubular surfaces improves the degree of enzymatic immobilization.

Conflict of interest

The authors declare no conflict of interest.

Author details

Crina Anastasescu, Mihai Anastasescu, Ioan Balint and Maria Zaharescu*
"Ilie Murgulescu" Institute of Physical Chemistry of the Romanian Academy, Bucharest, Romania

*Address all correspondence to: mzaharescu@icf.ro

References

[1] Shiraishi F, Toyoda K, Fukinbara S, Obuchi E, Nakano K. Photolytic and photocatalytic treatment of an aqueous solution containing microbial cells and organic compounds in an annular-flow reactor. Chemical Engineering Science. 1999;**54**(10):1547-1552. DOI: 10.1016/S0009-2509(99)00068-8

[2] Parvulescu V, Dascalescu C, Su BL. Highly selective oxidation of styrene with hydrogen peroxide catalyzed by mono-and bimetallic (Ni, Ni-Cr and Ni-Ru) incorporated MCM-41 silicas. Studies in Surface Science and Catalysis. 2003;**146**:629-632. DOI: 10.1016/S0167-2991(03)80462-7

[3] Nassif N, Bouvet O, Rager MN, Roux C, Coradin T, Livage J. Living bacteria in silica gels. Nature Materials. 2002;**1**:42-44. DOI: 10.1038/nmat709

[4] Mondal S, Malik S, Sarkar R, Roy D, Saha S, Mishra S, et al. Exuberant immobilization of urease on an inorganic support SiO_2 enhancing the enzymatic activities by threefold for perennial utilization. Bioconjugate Chemistry. 2019;**30**(1):134-147. DOI: 10.1021/acs.bioconjchem.8b00796

[5] Jiang Y, Wang W, Li X, Wang X, Zhou J, Mu X. Enzyme –mimetic catalyst –modified nanoporous SiO_2-cellulose hybrid composites with high specific surface area for rapid H2O2 detection. ACS Applied Materials and Interfaces. 2013;**5**(6):1913-1916. DOI: 10.1021/am400253d

[6] Kadam AA, Yang J, Lee DS. Supermagnetically tuned halloysite nanotubes functionalized with aminosilane for covalent laccase immobilization. ACS Applied Materials & Interfaces. 2017;**9**(18):15492-15501. DOI: 10.1021/acsami.7b02531

[7] Anastasescu C, Anastasescu M, Teodorescu VS, Gartner M, Zaharescu M. SiO_2 nanospheres and tubes obtained by sol–gel method. Journal of Non-Crystalline Solids. 2010;**356**:2634-2640. DOI: 10.1016/j.jnoncrysol.2010.03.038

[8] Anastasescu C, Mihaiu S, Preda S, Zaharescu M. 1D Oxide Nanostructures Obtained by Sol-Gel and Hydrothermal Methods. Cham, Switzerland: Springer; 2016. ISBN 978-3-319-32988-8

[9] Anastasescu C, Anastasescu M, Zaharescu M, Balint I. Platinum-modified SiO_2 with tubular morphology as efficient membrane-type microreactors for mineralization of formic acid. Journal of Nanoparticle Research. 2012;**14**:1198. DOI: 10.1007/s11051-012-1198-5

[10] Anastasescu C, Zaharescu M, Balint I. Unexpected photocatalytic activity of simple and platinum modified tubular SiO_2 for the oxidation of oxalic acid to CO2. Catalysis Letters. 2009;**132**:81-86. DOI: 10.1007/s10562-009-0066-0

[11] Spătaru T, Kondo T, Anastasescu C, Balint I, Osiceanu P, Munteanu C, et al. Silica veils-conductive diamond powder composite as a new propitious substrate for platinum electrocatalysts. Journal of Solid State Electrochemistry. 2017;**21**(4):1007-1014. DOI: 10.1007/s10008-016-3454-6

[12] Anastasescu C, Zaharescu M, Angelescu D, Munteanu C, Bratan V, Spataru T, et al. Defect-related light absorption, photoluminiscence and photocatalytic activity of SiO_2 with tubular morphology. Solar Energy Materials and Solar Cells. 2017;**159**:325-335. DOI: 10.1016/j.solmat.2016.09.032

[13] Spataru N, Anastasescu C, Radu MM, Balint I, Spataru T, Fujishima A. The improvement of SiO_2 nanotubes electrochemical behavior by hydrogen

atmosphere thermal treatment. Applied Surface Science. 2018;**444**:216-223. DOI: 10.1016/j.apsusc.2018.03.074

[14] Anastasescu C, Negrila C, Angelescu DG, Atkinson I, Anastasescu M, Spataru N, et al. Particularities of photocatalysis and formation of reactive oxygen species on insulators and semiconductors: Cases of SiO_2, TiO_2 and their composite SiO_2-TiO_2. Catalysis Science & Technology. 2018;**8**: 5657-5668. DOI: 10.1039/C8CY00991K

[15] Enache M, Neagu S, Anastasescu C, Cojoc R, Zaharescu M. The effects of silica nanostructures on halotolerant microorganisms isolated from rock salt crystal. New applications of nanomaterials, series In: Dascălu D, Cârjă G, Ciurea ML, Catrinel Ion A, editors. Micro and Nanoengineering. Ed Acad Rom, Bucharest: Romania; 2014. pp. 51-59

[16] Neagu S, Preda S, Anastasescu C, Zaharescu M, Enache M, Cojoc R. The functionalization of silica and titanate nanostructures with halotolerant proteases. Revue Roumaine de Chimie. 2014;**59**(2):97-103

[17] Anastasescu C, Preda S, Rusu A, Culita D, Plavan G, Strungaru S, et al. Tubular and spherical SiO_2 obtained by sol-gel method for lipase immobilization and enzymatic activity. Molecules. 2018;**23**:1362. DOI: 10.3390/molecules23061362

[18] Nakamura H, Matsui Y. Silica gel nanotubes obtained by the sol-gel method. Journal of the American Chemical Society. 1995;**117**:2651-2652. DOI: 10.1021/ja00114a031

[19] Miyaji F, Davis SA, Charmant JPH, Mann S. Organic crystal templating of hollow silica fibers. Chemistry of Materials. 1999;**11**(11):3021-3024. DOI: 10.1021/cm990449v

[20] Bai Y, Yang H, Yang W, Li Y, Sun C. Gold nanoparticles-mesoporous silica composite used as an enzyme immobilization matrix for amperometric glucose biosensor construction. Sensors and Actuators B. 2007;**124**:179-186. DOI: 10.1016/j.snb.2006.12.020

[21] Gustafsson H, Johansson EM, Barrabino A, Odén M, Holmberg K. Immobilization of lipase from Mucor miehei and Rhizopus oryzae into mesoporous silica - the effect of varied particle size and morphology. Colloids and Surfaces B: Biointerfaces. 2012;**100**:22-30. DOI: 10.1016/j.colsurfb.2012.04.042

[22] Mukhopadhyay K, Phadtare S, Vinod VP, Kumar A, Rao M, Chaudhari RV, et al. Gold nanoparticles assembled on amine-functionalized Na-Y zeolite: A biocompatible surface for enzyme immobilization. Langmuir. 2003;**19**(9):3858-3863. DOI: 10.1021/la0268202

[23] Nguyen DT, Smit M, Dunn B, Zink JI. Stabilization of creatine kinase encapsulated in silicate sol-gel materials and unusual temperature effects on its activity. Chemistry of Materials. 2002;**14**:4300-4306. DOI: 10.1021/cm020398t

[24] Ruscher CH, Bannat I, Feldhoff A, Ren L, Wark M. SiO_2 nanotubes with nanodispersed Pt in the walls. Microporous and Mesoporous Materials. 2007;**99**:30-36. DOI: 10.1016/j.micromeso.2006.07.032

[25] Lopez T, Romero A, Gomez R. Metal-support interaction in Pt/SiO_2 catalysts prepared by sol-gel method. Journal of Non-Crystalline Solids. 1991;**127**:105-113. DOI: 10.1016/0022-3093(91)90406-V

[26] Bey Temsamani M, Maeck M, El Hassani I, Hurwitz HD. Fourier transform infrared investigation of water states in Aerosol-OT reverse micelles as a function of counterionic

nature. The Journal of Physical Chemistry. B. 1998;**102**:3335-3340. DOI: 10.1021/jp971844g

[27] Pickup DM, Mountjoy G, Wallidge GW, Anderson R, Cole JM, Newport RJ, et al. A structural study of $(TiO_2)_x(SiO_2)_{1-x}$ (x=0.18, 0.30 and 0.41) xerogels prepared using acetylacetone. Journal of Materials Chemistry. 1999;**9**:1299-1305. DOI: 10.1039/a809810g

[28] Gustafsson H, Thorn C, Holmberg K. A comparison of lipase and trypsin encapsulated in mesoporous materials with varying pore sizes and pH conditions. Colloids and Surfaces B: Biointerfaces. 2011;**87**:464-471. DOI: 10.1016/j.colsurfb.2011.06.012

[29] Wang C, Zhou G, Xu Y, Chen J. Porcine pancreatic lipase immobilized in amino-functionalized short rod-shaped mesoporous silica prepared using poly(ethylene glycol) and triblock copolymer as templates. Journal of Physical Chemistry C. 2011;**115**:22191-22199. DOI: 10.1021/jp206836v

[30] Chen Y, Chen Q, Song L, Li HP, Hou FZ. Preparation and characterization of encapsulation of Europium complex into meso-structured silica monoliths using PEG as the template. Microporous and Mesoporous Materials. 2009;**122**:7-12. DOI: 10.1016/j.micromeso.2008.12.021

[31] Cao L. Immobilised enzymes: Science or art? Current Opinion in Chemical Biology. 2005;**9**:217-226. DOI: 10.1016/j.cbpa.2005.02.014

[32] Yu WH, Fang M, Tong DS, Shao P, Xu TN, Zhou CH. Immobilization of Candida rugosa lipase on hexagonal mesoporous silca and selective estrification in nonaqueous medium. Biochemical Engineering Journal. 2013;**70**:97-105. DOI: 10.1016/j.bej.2012.10.005

[33] Tran DT, Chen CL, Chang JS. Immobilization of Brukholderia sp. lipase on a ferric nanocomposite for biodiesel production. Journal of Biotechnology. 2012;**158**:112-119. DOI: 10.1016/j.jbiotec.2012.01.018

[34] Mendes AA, Oliveira PC, de Castro HF. Properties and biotechnological applications of porcine pancreatic lipase. Journal of Molecular Catalysis B: Enzymatic. 2012;**78**:119-134. DOI: 10.1016/j.molcatb.2012.03.004

[35] Stine KJ. Enzyme immobilization on nanoporous gold: A review. Biochemistry Insights. 2017;**10**:1-12. DOI: 10.1177/1178626417748607

[36] Du X, Liu X, Li Y, Wu C, Wang X, Xu P. Efficient biocatalyst by encapsulating lipase into nanoporous gold. Nanoscale Research Letters. 2013;**8**(1):180. DOI: 10.1186/1556-276X-8-180

Chapter 9

Electrospinning Technology: Designing Nanofibers toward Wound Healing Application

Daniela Sousa Coelho, Beatriz Veleirinho, Thaís Alberti, Amanda Maestri, Rosendo Yunes, Paulo Fernando Dias and Marcelo Maraschin

Abstract

Electrospinning is a widely used technology to obtain nanofibers. Electrospun systems have been especially investigated for wound dressings in skin regeneration given the similarity of structures with the extracellular matrix. Several efforts have been made to combine distinct design strategies, such as utilizing synthetic and/or natural materials, modifying fiber orientation, and incorporating substances, e.g., drugs, peptides, growth factors or other biomolecules, to develop an optimized electrospun wound dressing. This chapter reviews the current advances in electrospinning strategies for skin regeneration.

Keywords: electrospinning, nanofiber, wound healing, drug delivery system

1. Introduction

Skin is the essential interface between the body and its environment [1]. Besides providing a physical barrier that prevents pathologic infection, the skin also performs a range of vital functions that maintain hydration, thermoregulation, and body metabolism. Several injuries, such as burns and chronic wounds, result in lifelong functional impairment, and they represent a substantial burden on healthcare by the necessity of chronic medical care [2]. One promising approach to promote skin regeneration involves engineering of the local environment to promote coordinated cellular infiltration and organized deposition of extracellular matrix (ECM), and to provide a microenvironment prone to neodermis regeneration and appendage formation when combined with competent dermal cells. An ideal nanofibrous cell scaffold should resemble the native extracellular matrix and be capable of supporting cell adhesion, proliferation, and maturation [3, 4]. Nanofibrous scaffolds obtained by electrospinning have been attracting the attention of researchers owing to the morphological and structural similarities between the electrospun structures and the natural ECM, making these materials potential substrata for cell growth. Several electrospun scaffolds have been proposed for skin regeneration based on different polymers and their blends [5]. In this chapter, we present some recent advances in electrospun

IntechOpen

nanofibers for wound healing. We further highlight recent studies that have used electrospun nanofibers for wound healing applications and devices, including burns and nonhealing wounds.

2. Electrospinning nanofibers

Electrospinning (ES) is a technology to obtain nanomaterials formed by the deposition of polymer nanofibers, which results in an interconnected three-dimensional network [6]. Despite their extremely small diameter, these nanofibers show large surface area to volume ratio and high porosity. In addition, it is possible to obtain scaffolds from hundreds of different polymers capable of carrying bioactive substances [1]. Versatility is one of the most important advantages of the electrospinning technique. Different morphologies can be achieved by varying configuration and parameters of the electrospinning process [7]. Taken together, these characteristics show applicability in many different areas, such as high performance, intelligent textiles, biosensors, scaffolds for tissue engineering, and drug delivery systems [5]. A typical electrospinning apparatus includes a high voltage power supply, a spinneret system, and a collecting system (**Figure 1**). Tension, distance of the collector, and polymer solution flow rate are examples of parameters that could be changed. Creative electrospinning setups have been developed in order to obtain mats with distinctive characteristics, such as aligned, porous, hollow, and core-shell nanofibers [8]. Furthermore, many systematic studies have led to advances in knowledge about governing parameters of the electrospinning process. Nonetheless, optimizing the electrospinning system is still a laborious task as a result of the high number of parameters that affect the process and the interdependence among them. Even more, both the electrospinning setup and parameter optimization are tightly related to the polymer and solvent system [1]. In this section, some general principles about the parameters that affect the electrospinning process will be discussed, focusing on polymer characteristics that affect skin repair applications.

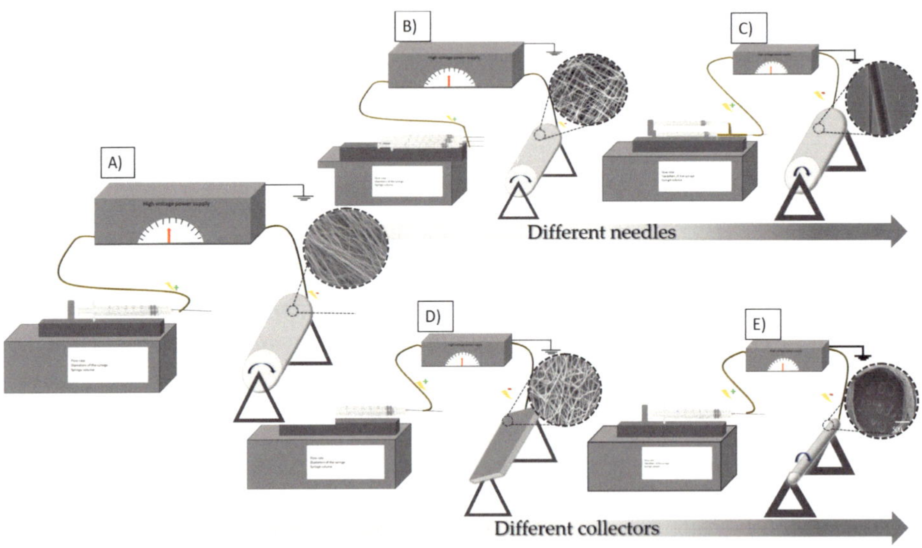

Figure 1.
Typical electrospinning apparatus equipped with different collectors and syringe systems. (A) Conventional electrospinning apparatus configured to produce aligned nanofibers (high-speed rotating drum). (B) Multijets for random nanofiber arrangement (low-speed rotating drum). (C) Co-axial electrospinning for obtaining core-shell nanofibers. (D) Static collector for random nanofibers arrangement. (E) Mandrel collector designed to produce nanofibrous tubes. The micrographs were adapted from De Prá and coworkers (2016).

2.1 Biomaterial characteristics

Materials considered as substitutes for skin repair should offer porosity to facilitate the clearing of exudates from the wound site, control water loss, and promote oxygen diffusion; hydrophilic surface to keep skin moist and moisturized; and controllable biodegradability to allow the controlled release of bioactive compounds that act on healing and suitable biocompatibility [9]. The chemical composition of the electrospinning solution determines the physicochemical, mechanical, and biological properties of the electrospun mat [10]. The intrinsic properties of the selected polymers are reflected in the final characteristics of the electrospun mat, while the nanofibrous arrangement brings unique features to the material.

A polymer consists of a long chain of molecules with repeating units called monomers that are mostly covalently bonded to one another [11]. Varying concentration and molecular weight (MW) of the polymer are the most effective ways to control the morphology of the electrospun mats, especially the diameter of the fibers [12]. Molecular weight (MW) is a key determinant of electrospinning processability of any polymer solution by strongly affecting its rheological and electrical properties, including viscosity, surface tension, conductivity, and dielectric strength [2]. Generally, low MW polymer solutions are more suitable for electrospray purposes, i.e., formation of beads, while high MW polymer solutions possess the desired viscosity, i.e., sufficient polymer chain entanglements to generate fibers [13]. Katti et al. demonstrated the effect of poly(lactide-co-glycolide) (PLGA) solution concentration on the diameter and morphology of fibers. Nanofibers were formed for intermediate polymer concentrations, while for low polymer concentration (0.10 g/mL), no fibers were formed, only beads. The highest concentration (0.30 g/mL) produced fibers with thickness in the microns [14]. Similar results were found with poly(ethylene terephthalate) (PET) for concentrations lower than 10% (w/v). In this case, droplet spray occurred, and a continuous jet of polymer was not formed. For concentrations higher than 30% (w/v), the high viscosity of the solution limited Taylor cone formation, the cone observed in electrospinning, electrospraying and hydrodynamic spray processes from which a jet of charged particles emanates above a threshold voltage [12].

Recently, many researchers have been using electrospinning technology to develop new scaffolds made of natural polymers, such as collagen, gelatin, chitosan, and silk fibroin, as well as synthetic biodegradable polymers [15]. Natural polymers have been found to promote cell attachment and proliferation, whereas synthetic polymers provide mechanical stability to the scaffold [16].

2.2 Collagen

Collagen is the major fibrous protein of the ECM. It constitutes 20–30% of total body protein, plays an important role in the regulation of cell function, and provides structural support for tissues and organs [17]. Biocompatibility is the most favorable aspect of using collagen as biomaterial for nanofiber mats. As a result of its abundance in the body, it is easily available. Furthermore, while it is nontoxic and not antimitogenic, it is biodegradable by enzymes, which naturally occur in ECM substitution during the remodeling phase of the healing process. It is also possible to combine collagen with other copolymers since it is highly compatible and has good mechanical properties, such as malleability and bioresorbability [18]. However, the main reason against using collagen is the onerous work involved and high cost of its purification, as well as the risk of disease transmission.

Gelatin is the denatured form of collagen and has different physicochemical and biological characteristics compared to natural collagen [19]. A comparative study between electrospun collagen and gelatin revealed that collagen induces a better cellular response [20].

2.3 Chitosan

Chitosan is a cationic polysaccharide and has been one of the most studied biopolymers in the biomedical field owing to its notable proprieties, i.e., biocompatibility and wound healing effect, as well as anti-inflammatory and antimicrobial activity [21, 22]. However, the poor mechanical properties of chitosan pristine fibers have restricted biomedical applications, even though some studies have reported good results for chitosan composite fibers, especially when combined with synthetic polymers [5, 22–25]. Nanofibrous mats electrospun with chitosan-graft-polycaprolactone have shown excellent properties in cell attachment and proliferation, and they are promising substitutes for skin tissue engineering [5, 24].

2.4 Silk fibroin

Silk is a typical fibrous protein produced by a variety of insects, including silkworm. Silk consists of two types of proteins: fibroin and sericin. Fibroin is the protein that forms the filaments of silkworm silk, and it can be regenerated in various forms, such as gels, powders, fibers, or membranes, depending on the application [26]. Among silk proteins, silk fibroin (SF) has recently been investigated as one of the candidate materials for biomedical applications based on its several distinctive biological properties, including good biocompatibility, biodegradability, minimal inflammatory reaction, and suitable oxygen and water vapor permeability [27].

2.5 Poly(ε-caprolactone)

Polycaprolactone (PCL) is a biodegradable and biocompatible poly(alfa-ester) and one of the polymers that has been extensively studied in tissue regeneration and wound healing applications because it promotes faster healing and reduced inflammatory infiltrate [28, 29]. PCL has some important physicochemical properties, including hydrophobicity, excellent spin ability, favorable mechanical properties, and slow degradation, all of which support its use as a good matrix for loading natural substances.

2.6 Polyvinyl alcohol (PVA)

PVA is a semicrystalline polymer that shows excellent electrospinnability in aqueous solution [30]. Furthermore, PVA has been used as a copolymer to enhance electrospinnability and mechanical properties of biopolymers like chitosan [31, 32]. The biocompatibility of PVA allows its application in biomaterials for dermal and orthopedic tissue engineering [7, 31]. PVA was associated with chitosan blends in which the increase of chitosan derivative ratio resulted in smaller fiber diameters and higher antibacterial effect, both advantageous properties against skin infection [23].

In this section, we describe polymer solution parameters and their resultant application in the electrospinning process. However, it should be noted that other parameters related to the process also affect fiber morphology obtained as a result of electrospinning. Some process parameters include the applied electric field, tip-to-collector distance, and flow rate [33].

Some parameters and their effects on fiber morphology are summarized in **Table 1**.

Parameters	Effects on fiber morphology	References
Applied voltage	Decrease in fiber diameter with increase in voltage	[37, 47, 48]
Type of collector	Decrease of fiber diameter and increase in fiber alignment with rotating drum (compared to static collector)	[34, 38]
Distance of collector	Generation of beads with too long or too short distance; minimum distance required for uniform fibers	[39, 40, 42, 49]
Flow rate	Decrease in fiber diameter with decrease in flow rate; generation of beads with too high flow rate	[41, 43]
Air humidity	High air humidity results in circular pores on the fiber surface	[44]
Temperature	Increase in temperature causes decrease in fiber diameter	[46]

Adapted from Bhardwaj and Kundu [34].

Table 1.
Electrospinning process parameters and their effects on fiber morphology.

2.7 Effect of applied voltage

A crucial element in electrospinning is the application of high voltage to the solution. According to Ramakrishna et al., voltage of more than 6 kV can cause the solution drop at the tip of the needle to distort into the shape of a Taylor Cone during Jet initiation [35]. If the applied voltage is higher, the greater amount of charges will cause the jet to accelerate faster, and more volume of solution will be drawn from the tip of the needle. Thus, the formation of beads or beaded nanofiber can be attributed to an increase in the applied voltage [35]. Matabola and Moutloali evaluated the influence of applied voltage on the morphology and diameter of poly(vinylidene fluoride) (PVDF) nanofiber and showed that the diameter increased gradually with decreasing bead density at lower voltages, whereas at higher voltage, the reverse situation was observed whereby diameters decreased and beads reappeared [36]. In another study, it was observed that increasing voltage favors the formation of multiple jets. The different fibers repel each other owing to the flowing charge on their surface, and as a result, the fibers distribute themselves at a larger area on the collector [37].

2.8 Effect of collector

For electrospinning to initiate, an electric field must be present between the source and the collector. Thus, the collector plate is made out of conductive material. If a nonconducting material is used as a collector, charges on the electrospinning jet will quickly accumulate on the collector. Fibers collected on the nonconducting material usually have lower packing density compared to those collected on a conducting surface [34].

The electrospinning process is also affected by the use of a static or moving collector. While rotating collectors have been used to collect aligned fibers, static models randomly arrange nanofibers in the collector. Our group discovered that three distinct metallic collectors, including a rotating drum, 6 mm grounded parallel copper wires, and 1 mm rotating mandrel, had an effect on PCL nanofiber morphology. For the static drum, randomly oriented fibers with an average diameter were obtained (**Figure 1(D)**). Increasing the rotation speed to 2000 rpm caused a significant decrease in the mean fiber diameter (**Figure 1(A)**). A similar fiber alignment pattern was obtained using parallel copper. For the rotating mandrel, tubular scaffold formation was fundamental for engineering of nerves and blood vessels (**Figure 1(E)**) [38].

2.9 Distance between tip and collector

Varying the distance between the tip and the collector will have a direct influence on flight time and electric field strength [39–41]. If the distance is so short that the solvent is inadequately vaporized, then fused fiber may be formed. Depending on the solution property, the effect of varying the distance may or may not have a significant effect on fiber morphology. Ki et al. showed four series of spinning distance under a fixed electrical field. The results were not significant; however, many droplets on the electrospun gelatin web were observed at farther distance [42].

2.10 Flow rate

The flow will determine the amount of solution available for electrospinning. A lower flow rate is more favorable as the solvent will have more time for evaporation. In contrast, if a greater volume of solution is drawn from the needle tip, the jet will take a longer time to dry. As a result, the solvents in the deposited fiber may not have enough time to evaporate. In this case, residual solvent remaining in the collector may cause fusion between the fibers, forming unwanted webs [41, 43].

2.11 Air humidity

The humidity of the electrospinning environment may influence the polymer solution during electrospinning. Under high humidity, it is likely that water will condense on the surface of the fiber, which could have an effect on fiber morphology, especially polymer dissolved in volatile solvents [44]. However, it is possible to use humidity to develop porous electrospinning. Bae et al. explain how porous electrospun polymethyl methacrylate (PMMA) can be used as water filters to optimize humidification. Increased humidity will increase the amount of porosity. In this work, the PMMA fiber membrane obtained at 70% humidity was highly porous, compared to that at 25% humidity, but with no differences in the mean pore diameter [45].

2.12 Temperature

Increasing the temperature has two effects on the polymeric solution: increasing the evaporation rate of the solvent and reducing the viscosity of the polymer solution. As a result, the nanofibers show a decrease in diameter of the fiber and more homogeneous distribution. This will improve the solubility of the polymer in the solution and, hence, the attraction of polymeric particles to the collector. On the other hand, when the matrix is developed for biomedical application, biological molecules are added, and these substances can be degraded by high temperature, causing loss in functionality [33, 46].

3. Innovative electrospinning techniques

Despite the numerous advantages offered by nanofiber, the development of new composite nanofibers holds even greater potential, and investigation of new design and synthesis of composite materials further expands their applicability [50–52]. Consequently, nanofiber composite fabrication using electrospinning techniques has gained attention in recent years. Fiber composites obtained by blending multicomponent polymer mixtures or by dispersing nanofillers within fibrous matrix are promising as these systems can have excellent optical, electrical, or magnetic properties, making it easier to produce functional fibers [10]. For example, next-generation wound dressing materials should not only prevent pathogens from entering the wound, but should also be capable of monitoring the status of the wound, aiding the healing process and delivering drugs directly to the wound area [53]. A diverse series of methodologies have been used to fabricate composite nanofibers. The biocompounds can be loaded to the mat by different methods, including coelectrospinning, side-by-side, multijet, coaxial, emulsion, and surface immobilization [9]. By coelectrospinning, the drug could be homogenized with the polymer solution. This tactic is valid when the compounds are stable and soluble in the same solvent as the polymer; however, the bioactivity of drug molecules may be affected as a result of interaction between solvents and the electric field [54]. Another problem commonly faced occurs when the biocompounds are insoluble in a common solvent. To solve the solubility issue, side-by-side electrospinning provides two solvent spinnerets at the same time [55]. Also, more than two spinnerets can be used to load different molecules by multijet electrospinning (**Figure 1B**). Coaxial electrospinning involves fabrication of nanofibers from two polymers from a coaxial capillary spinneret, and as a result, the core and the shell are formed by different polymers (**Figure 1C**) [33]. With this technology, some polymers which are difficult to process are coelectrospun and form a core inside the shell of another polymer. Electrospun nanofibers are also used as drug delivery vehicles, but because of their large surface area and high porosity, a significant burst release is frequently observed. The coaxial method is used to control the burst release of drugs as the shell of the polymer acts as a diffusion barrier for drugs [56]. Emulsion electrospinning can produce a drug/polymer nanofiber core-shell structure, which possesses an excellent ability to control the release rate of the drug and avoid the initial burst. In the surface immobilization method, the drug molecules can be covalently bonded to the surface of nanofibers by physical and chemical immobilization. This method can also protect the bioactivity of loaded molecules from the effects of high voltage [10].

4. Nanofibers and biomedical application

Nanofibers have been widely used in various biomedical applications, including drug delivery [57], tissue engineering [53], stem cell therapy [34], cancer therapy [13, 57], and wound healing [58–60]. This is because they offer numerous attractive features, such as large surface area, material design flexibility, and tunable functional properties, which facilitate and widen the use of nanofibers in a variety of biomedical applications [50, 61]. The following section describes the recent advances in the use of nanofibers for drug delivery and wound healing applications.

4.1 Applications of nanofiber in skin wound healing

As mentioned above, nanofibrous scaffolds seem to be a good candidate as a skin substitute for wound healing, especially by their similarity to ECM. The ECM is a complex structure that surrounds cells in all tissues of the body [62]. Its biochemical composition varies somewhat from tissue to tissue. In healthy skin, the ECM consists of fibrous structural proteins, including collagens, elastins, laminins, and a variety of polysaccharides and proteoglycans, e.g., dermatan sulfate and hyaluronan [1]. Furthermore, it helps to support cells and comprises key components of the basement membrane that anchor and help replenish epidermal cells [54]. An excellent work published by Felgueiras and Amorim cited a previous review about acute and chronic wounds. Acute wound healing is a well-organized process, following four phases known as hemostasis, inflammation, proliferation, and remodeling (**Figure 2**). Immediately after microvascular injury and extravasation of blood and its components into the wound, a complex process involving coordinated interaction between diverse immunological and biological systems begins [7, 10]. Along with hemostatic events, the coagulation cascade is activated through extrinsic and intrinsic pathways, leading to platelet aggregation and clot formation in order to limit blood loss (**Figure 2B**) [29]. The fibrin clot also has an important function in signaling further events of the healing process since it elicits numerous cytokines and growth factors. The degranulation of platelets triggers the recruitment of inflammatory cells to clean the wound. This is the beginning of the inflammatory phase, which can be macroscopically identified by redness, heat swelling, and pain. Vasodilation and increased permeability of blood vessels occur, facilitating the penetration of plasma proteins and leucocytes into the wound area (**Figure 2C**). Macrophages and neutrophils are the first cells to arrive at the wound site. The main role of these cells is to eliminate bacteria, debris, and foreign bodies. Despite the rapid action of neutrophils, the activity of these cells is strictly associated with damage in the surrounding tissues. During the proliferative phase, fibroblasts are activated, proliferate, and start to produce and secrete collagen, which replaces the provisional ECM. The granulation tissue is characterized by a high density of new blood vessels as a result of intense angiogenesis (**Figure 2D**). Re-epithelialization occurs as a result of keratinocyte proliferation and migration from the wound edges and skin appendages toward the center of the wound. Remodeling of tissue initiates during the proliferative phase and persists for weeks, months, or years. During remodeling, the cellularity of the granulation tissue gradually decreases. Additionally, a series of rearrangements of ECM take place (**Figure 2E**) [7]. Unlike acute wounds, chronic wounds are a result of gradual tissue degradation by chemical and biological agents like proteolytic enzymes derived from neutrophils. In chronic wounds, the level of some proteases increased ten- to fortyfold over that in acute wounds. This proteolytic activity may lead to a continued degradation of the tissue and stop the healing process [63]. The main factors that contribute to the nonhealing condition of a wound include infection, advanced age, diabetes, and other chronic diseases [64], as well as high levels of metalloproteinases and lack of the integrin receptor for fibronectin binding and keratinocyte migration [65]. In the past, traditional dressings were used to simply manage wounds, including

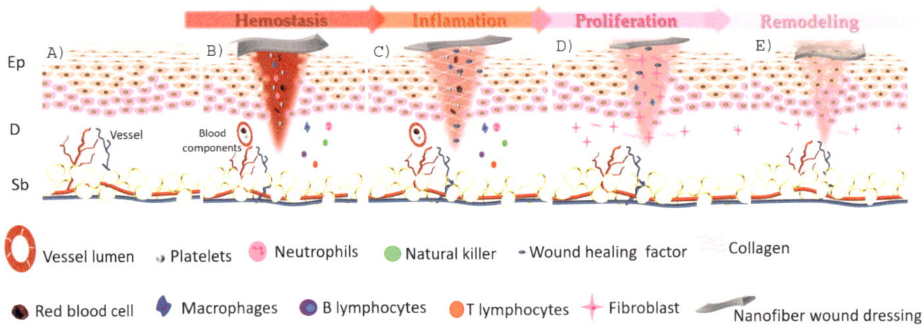

Figure 2.
Wound dressing nanofiber. Layers of epidermis (Ep), dermis (D), and subcutaneous tissue (Sc). Immediately following cutaneous injury, blood components (neutrophils, macrophages, platelets, and system immune cell) extravasate from the dermis and infiltrate the wound. The coagulation occurs as platelets aggregate with fibrin, which is deposited in the wound following its conversion in fibrinogen. Platelets release several factors, which attract neutrophils for the local injury, signaling the beginning of inflammation phase. Neutrophils and macrophages remove debris from the wound, release growth factor, which recruit fibroblast to the wound and start to synthesize collagen in the proliferation phase. As the rate of collagen synthesis slows down, the reorganization occurs in the remodeling phase. Throughout all stages, the nanofiber composite interacts with the tissue, releasing drugs that will facilitate the healing process. This figure is an adaptation from Alberti and coworkers [7].

gauze, lint, plaster, bandages, and cotton wool. Care included the use of antibiotics, medicated ointments and creams, steroids, vitamin injection, and laser therapy. Modern dressing technology has evolved considerably in the last 50 years. It is now based on the principle of creating and maintaining a moist wound environment that stimulates proliferation and migration of fibroblast and keratinocytes and enhances collagen synthesis [7, 63]. As a result, creating strategies to replace the missing or dysfunctional ECM may be a major goal of skin engineering, aiming to help the body to complete regeneration when a given clinical condition hampers the normal healing process. A common approach has been the use of biodegradable scaffolds to sustain and guide cellular growth throughout the regeneration process [66]. The scaffold's role is to provide an artificial environment for cell adhesion, migration, and spread, thus allowing cell proliferation. Furthermore, an ideal dressing scaffold should have certain characteristics, such as hemostatic ability; efficiency as a bacterial barrier; ability to absorb excess exudates (wound fluid/pus); exhibit appropriate water vapor transmission rate; exhibit adequate gaseous exchange ability; ability to conform to the contour of the wound area; exhibit functional adhesion, i.e., be adherent to healthy tissue, but nonadherent to wound tissue; be painless to the patient with ease of removal and, finally, be available at low cost [67]. Over the next few years, it is predicted that many materials with temporary therapeutic application will be substituted for biodegradable biomaterials with an active role in regeneration and repair of damaged tissues [7]. The marketplace offers a range of materials for wound dressing. For example, collagen-based Integra® sponges have been used as temporary coverage to promote granulation tissue formation prior to autografting for extensive skin defects [1]. Integra™ (Integra Life Sciences) is composed of an outer silicone epidermis and a dermis composed of bovine tendon collagen and chondroitin-6-sulfate [68]. A few other substitutes are Laserskin™, Biomembrane™, Bioseed™, and Hyalograft3-DTM [69]. These commercial mats were used in a clinical case report, and the results showed complete healing and an acceptable functional and cosmetic outcome with minimal morbidity to the patients [70]. Electrospinning scaffolds also serve as promising substrates for tissue repair owing to their nanofibrous architecture and amenability to tailoring of chemical composition. After modification with gelatin, the scaffolds improved human dermal fibroblast infiltration and proliferation throughout the scaffolds and the secretion of ECM proteins from the cells, showing potential for dermal tissue engineering [71].

4.2 Electrospun scaffolds for skin engineering

Engineering mats for wound healing requires not only the appropriate processing, solution, and environmental conditions, but also selection of the most suitable polymer matrix or rational combination of polymers [72]. Several electrospun materials have been proposed as wound dressing materials, including natural polymer, such as polysaccharides [73], polymeric carbohydrate molecules [74], chitin [75], chitosan [23], alginates [46], cellulose [4], and hyaluronic acid [76]. These natural polymers can be used alone or in combination with synthetic polymer, such as polyurethane [37], PVA/Ag [77], PCL [78], such as collagen/chitosan [79], chitosan/polyethylene glycol (PEG) [80], and water-soluble carboxyethyl chitosan/PVA [69, 81]. Additionally, electrospun mats can be seeded with dermal and epidermal cell lines. Bonville and coworker reported that collagen and PCL electrospun scaffold with dermal fibroblast promoted greater wound healing than acellular scaffolds [82]. Electrospinning scaffolds have also been used for controlled release of drugs [14], herbal extracts [83–85], growth factors, such as FGF [86], EGF [87], and angiogenesis factors [88], has shown successful improvement in the wound healing process, including *in vivo* experiments. Also, the incorporation of growth factors, such as FGF [86] and EGF [87], or angiogenesis factors [88], has also been achieved and has shown successful improvement in the wound healing process, including *in vivo* experiments.

4.3 Drug delivery systems

Nanofiber composite-incorporated drugs play an important role in the healing process. This has been achieved by cleaning or debriding agents for necrotic tissue and antimicrobials, which prevent infection and promote tissue regeneration [51]. Some commonly incorporated compounds include antimicrobial agents, growth factors, and enzymes [69]. The main goal of drug delivery systems (DDS) is to efficiently deliver drug molecules within the recommended therapeutic level to the target cell, tissue, or organ for a defined period time [34]. A number of researchers have successfully encapsulated drugs within electrospun fibers by mixing the drugs in the polymer solution to be electrospun [7, 89, 90]. In most cases, polymer nanofibers are being used as candidate vehicles to carry the drug molecules. Depending on the polymer used, the release of a given pharmaceutical compound can be designed to have immediate, rapid, or delayed delivery [2]. This different profile of drug release is made possible by nanofiber biodegradability and can be adjusted according to the affinity of drug and mats to the target tissue. Commercially available antimicrobial dressings include Cutis orb™. Available silver impregnated dressings include fibrous hydrocolloid, polyurethane foam film, and silicone gels. Antiseptic iodine dressing acts on bacterial cells via oxidative degradation of cell components by interrupting the function of protein, which is widely effective against pathogens. Prolonged usage of iodine leads to skin irritation and staining [91]. Electrospun nanofibrous scaffolds constitute an excellent platform for local delivery of therapeutic agents, e.g., antimicrobial agents, antioxidants, anti-inflammatory drugs, anesthetics, enzymes, and growth factors, which have been incorporated into electrospun nanofibers for wound healing purposes [1]. Structural control of electrospun nanofibrous mats and loading suitable drugs are both important considerations in producing an active wound dressing material [92]. Therapeutic agents may be hydrophilic, hydrophobic, or both. Compounds such as vitamins, antibiotics, growth factors, anti-inflammatory agents, and other materials have been used. One of the most widely used additives is the nanoAg particle, which is incorporated into biopolymeric electrospun membranes by different methods. For example, Rujitanaroj and coworkers developed mats made of gelatin fibers containing silver nanoparticles lastly; the antibacterial activity of these materials was greatest against *Pseudomonas aeruginosa*, followed by *Staphylococcus aureus*, *Escherichia coli*, and methicillin-resistant *S. aureus* [93]. In another study, a mixture of PVA and chitosan oligosaccharides (COS) was electrospun with silver nanoparticles to produce fibrous mats for use in wound healing. *In vivo* wound healing experiments suggest that PVA/COS-AgNPs nanofibers can have clinical applications as a bioactive wound dressing [94]. Zhang et al. recently reported electrospun silver (Ag) nanoparticle-embedded polyvinyl alcohol (PVA) nanofibers as an antimicrobial scaffold. Silver nanoparticles (NPs) were first prepared and dispersed in PVA solution, these nanofiber mats showed good antibacterial activity against both gram-positive *S. aureus* and gram-negative *Escherichia coli* microorganisms [95]. Barani also reported the fabrication of an antibacterial nanofibrous mesh from PVA and poly-L-lactide acid (PLLA) with AgNPs [96]. For wound healing purposes, two specific requirements for a wound scaffold include rapid hemostasis and good antibacterial properties. Thus, antibacterial nanofibers have been fabricated by incorporating antibacterial materials into the polymers [67]. Beyond silver nanoparticles, other natural compounds have been associated with nanofibers for

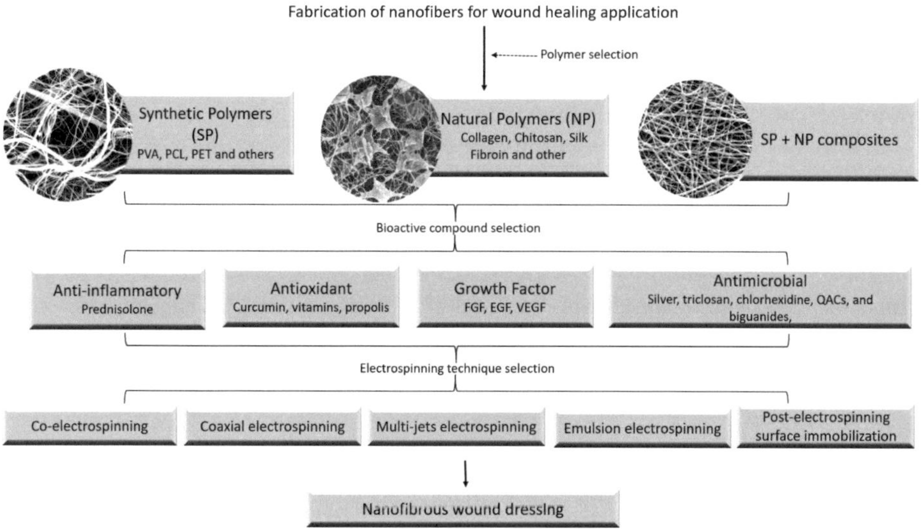

Figure 3.
Schematic illustration of the development of a nanofibrous wound dressing material.

antimicrobial activity, such as *Aloe Vera* [83], propolis [84], curcumin [97], and chitosan [66, 79]. Common antibacterial materials, such as antibiotics, triclosan, chlorhexidine, quaternary ammonium compounds (QACs), and biguanides, have also been reported for use in the fabrication of antibacterial nanofibers [3]. Silk/fibroin nanofibers were also functionalized with a sulfate group in order to test whether they exhibited antibacterial activity [60, 84, 90]. A schematic illustration of the main steps for the development of a nanofibrous wound dressing material is displayed in **Figure 3**.

5. Conclusion and outlook

Electrospinning has attracted the interest of researchers from different fields as a simple and versatile technique to produce nanofibers with a range of clinical applications. Moreover, because of enormous research progress over past decades, the development of electrospun scaffolds for skin engineering seems to be in the laboratory phase. Many studies have improved the knowledge of the electrospinning process, and different approaches have confirmed the potential of using electrospinning in wound healing applications, as demonstrated in this chapter. However, preclinical studies with promising results are needed before these products reach the marketplace. Furthermore, the transition of nanofiber technology remains confined owing to difficulties associated with upscaling the electrospinning process itself, such as control of pore size and elimination of artifacts like beads, as well as consistency in physical characteristics like porosity and distribution of nanofibers with homogeneous diameters. In conclusion, this chapter reports the main considerations for developing scaffolds that aid in the wound healing process. By developing nanofibers for wound dressing application, the selection of some parameters are crucial, including, for example, the selection of polymers, type of collector, and better systems for drug delivery in the local wound in order to provide low cytotoxicity, good biodegradation rate, permeability, avoidance of bacterial proliferation, provision of re-epithelization of the injured site, and promotion of keratinocyte and fibroblast growth since these cells act as an adjuvant in the wound healing process. Electrospun mats for skin tissue engineering should serve as a barrier for the wound site and possess a structure with high porosity, oxygen permeability, wettability, suitable mechanical properties, antimicrobial activity, surgical handleability, biodegradability, cell adhesive properties, and wound healing properties.

Electrospun nanofiber scaffolds incorporated with cells and/or bioactive compounds are a promising approach for skin tissue engineering, facilitating treatment and promoting wound healing of many skin disorders. However, the risks of working in this field have still not been assessed, including, for example, the inhalation of nanosized fibers and solvent vapor, which could very well pose health hazards. Thus, the risk-benefit ratio of this technology toward realizing its therapeutic potential warrants an interdisciplinary approach.

Acknowledgements

This work was supported by Conselho Nacional de Desenvolvimento Científico e Tecnológico (CNPq) (grant 445401/2014-1), CNPq (grant 454572/2014-0) and Coordenação de Aperfeiçoamento de pessoal de Nível Superior (CAPES). The authors also thank these agencies for their financing and fellowships.

Abbreviations

Ag	silver
$AgNO_3$	silver nitrate
AgNPs	silver nanoparticles
COS	chitosan oligosaccharides
DDS	drug delivery systems
ECM	extracellular matrix
EGF	epidermal growth factor
ES	electrospinning
FGF	fibroblast growth factor
MW	molecular weight
NPs	nanoparticles
PCL	poly(ε-caprolactone)
PEG	polyethylene glycol
PET	poly(ethylene terephthalate)
PLGA	poly(lactide-co-glycolide)
PLLA	poly-L-lactide acid
PMMA	polymethyl methacrylate
PVA	polyvinyl alcohol
PVDF	poly(vinylidene fluoride)
QACs	quaternary ammonium compounds
SF	silk fibroin

Author details

Daniela Sousa Coelho[1]*, Beatriz Veleirinho[2], Thaís Alberti[1], Amanda Maestri[1], Rosendo Yunes[3], Paulo Fernando Dias[1] and Marcelo Maraschin[1]

1 Federal University of Santa Catarina, Florianopolis, Brazil

2 NanoScoping—Soluções em Nanotecnologia, Florianopolis, Brazil

3 University of Itajaí Valley, Itajaí, Brazil

*Address all correspondence to: daniela.sousacoelho@gmail.com

References

[1] Chen R. Recent advances in electrospun nanofibers for wound healing. Nanomedicine (London, England). 2017;**12**:1335-1352

[2] Dias JR, Granja PL, Bártolo PJ. Progress in materials science advances in electrospun skin substitutes. Progress in Materials Science. 2016;**84**:314-334

[3] Anjum F, Agabalyan NA, Sparks HD, Rosin NL, Kallos MS. Biocomposite nanofiber matrices to support ECM remodeling by human dermal progenitors and enhanced wound closure. Scientific Reports. 2017;7:1-17

[4] Liu R, Dai L, Si C, Zeng Z. Antibacterial and hemostatic hydrogel via nanocomposite from cellulose nano fi bers. Carbohydrate Polymers. 2018;**195**:63-70

[5] Veleirinho B et al. Manipulation of chemical composition and architecture of non-biodegradable poly(ethylene terephthalate)/chitosan fibrous scaffolds and their effects on L929 cell behavior. Materials Science and Engineering: C. 2013;**33**:37-46

[6] Alfaro De-Prá, matrizes nanoestruturadas bioativas para a aplicação na regeneração de nervos periféricos. 2017;**1**:1-138

[7] Alberti T et al. Nanotechnology: A promising tool towards wound healing. Current Pharmaceutical Design. 2017;**23**:1-14

[8] Passalacqua TG et al. Synthesis and evaluation of novel prenylated chalcone derivatives as anti-leishmanial and anti-trypanosomal compounds. Bioorganic & Medicinal Chemistry Letters. 2015;**25**:3564-3568

[9] Alfaro De Prá MA, Ribeiro-do-Valle RM, Maraschin M, Veleirinho B. Effect of collector design on the morphological properties of polycaprolactone electrospun fibers. Materials Letters. 2017;**193**:154-157

[10] Liu M, Duan X, Li Y, Yang D, Long Y. Electrospun nano fibers for wound healing. 2017;**76**:1413-1423

[11] Baji A, Mai Y-W. Polymer nanofiber composites. 2017;**3**:55-78

[12] Ramakrishna S. An Introduction to Electrospinning and Nanofibers. World Scientific; 2005;**1**:1-382

[13] Veleirinho B, Rei MF, Lopes-Da-Silva JA. Solvent and concentration effects on the properties of electrospun polyethylene terephthalate nanofiber mats. Journal of Polymer Science Part B: Polymer Physics. 2008;**5**:460-471

[14] Thakkar S, Misra M. Electrospun polymeric nanofibers: New horizons in drug delivery. European Journal of Pharmaceutical Sciences. 2017;**10**:1393-1399

[15] Katti DS, Robinson KW, Ko FK, Laurencin CT. Bioresorbable nanofiber-based systems for wound healing and drug delivery: Optimization of fabrication parameters. Journal of Biomedical Materials Research. 2004. DOI: 10.1002/jbm.b.30041

[16] Huang S, Fu X. Naturally derived materials-based cell and drug delivery systems in skin regeneration. Journal of Controlled Release. 2010;**142**:149-159

[17] Ghasemi-Mobarakeh L, Prabhakaran MP, Morshed M, Nasr-Esfahani MH, Ramakrishna S. Electrospun poly(ε-caprolactone)/gelatin nanofibrous scaffolds for nerve tissue engineering. Biomaterials. 2008;**29**:4532-4539

[18] Zeugolis DI, Li B, Lareu RR, Chan CK, Raghunath M. Collagen solubility

testing, a quality assurance step for reproducible electro-spun nano-fibre fabrication. A technical note. Journal of Biomaterials Science. Polymer Edition. 2008;**19**:1307-1317

[19] Law JX, Liau LL, Saim A, Yang Y, Idrus R. Electrospun collagen nanofibers and their applications in skin tissue engineering. Tissue Engineering & Regenerative Medicine. 2017;**14**:699-718

[20] Maleknia L, Rezazadeh Majdi Z. Electrospinning of gelatin nanofiber for biomedical application. Oriental Journal of Chemistry. 2014;**30**:2043-2048

[21] Simpson DG et al. Electrospun collagen: A tissue engineering scaffold with unique functional properties in a wide variety of applications. Journal of Nanomaterials. 2011

[22] Lopes da Silva J, Veleirinho B, Delgadillo I. Preparation and characterization of electrospun mats made of PET/chitosan hybrid nanofibers. Journal of Nanoscience and Nanotechnology. 2009;**9**:3798-3804

[23] Wang M, Roy AK, Webster TJ, Webster T. Development of chitosan/poly (vinyl alcohol) electrospun nanofibers for infection related wound healing. Npg Asia Materials. 2017;7:2016-2018

[24] Alavarse AC et al. Tetracycline hydrochloride-loaded electrospun nanofibers mats based on PVA and chitosan for wound dressing. Materials Science and Engineering: C. 2017;**77**:271-281

[25] Chen H, Huang J, Yu J, Liu S, Gu P. Electrospun chitosan-graft-poly(ε-caprolactone)/poly(ε-caprolactone) cationic nanofibrous mats as potential scaffolds for skin tissue engineering. International Journal of Biological Macromolecules. 2011;**48**:13-19

[26] Ribeiro TG et al. An optimized nanoparticle delivery system based on chitosan and chondroitin sulfate molecules reduces the toxicity of amphotericin B and is effective in treating tegumentary leishmaniasis. International Journal of Nanomedicine. 2014;**9**:5341-5353

[27] Kaplan D, Adams WW, Farmer B, Viney C. Chapter 1. Silk: Biology, structure, properties, and genetics. 1994;**1**:2-16

[28] Rajkhowa R et al. Synthesis, characterization and efficacy of chemically crosslinked PVA hydrogel for dermal wound healing in experiment et al animals. Journal of Applied Polymer Science. 2008;**111**:1400-1408

[29] Moura LIF, Dias AMA, Carvalho E, de Sousa HC. Recent advances on the development of wound dressings for diabetic foot ulcer treatment—A review. Acta Biomaterialia. 2013;**9**:7093-7114

[30] Dash TK, Konkimalla VB. Poly-ϵ-caprolactone based formulations for drug delivery and tissue engineering: A review. Journal of Controlled Release. 2012;**158**:15-33

[31] Koski A, Yim K, Shivkumar S. Effect of molecular weight on fibrous PVA produced by electrospinning. Materials Letters. 2004;**58**:493-497

[32] Lin T, Fang J, Wang H, Cheng T, Wang X. Using chitosan as a thickener for electrospinning dilute PVA solutions to improve fibre uniformity. Nanotechnology. 2006;**17**:3718-3723

[33] Gupta A, Kumar R, Upadhyay N, Sureka P, Roy P. Synthesis, characterization and efficacy of chemically crosslinked PVA hydrogels for dermal wound healing in experimental animals. Journal of Applied Polymer Science. 2008;**111**:1400-1408

[34] Bhardwaj N, Kundu SC. Electrospinning: A fascinating fiber

fabrication technique. Biotechnology Advances. 2010;**28**:325-347

[35] Reddy J, Venugopal et al. Nanofibrous structured biomimetic strategies for skin tissue regeneration. Wound Repair and regeneration. 2013;**21**:1-16

[36] Hairder A, Sajjad Haider I-KK. A comprehensive review summarizing the effect of electrospinning parameters and potential applications of nanofibers in biomedical and biotechnology. Arabian Journal of Chemistry. 2015

[37] Matabola KP, Moutloali RM. The influence of electrospinning parameters on the morphology and diameter of poly(vinyledene fluoride) nanofibers- effect of sodium chloride. Journal of Materials Science. 2013;**48**:5475-5482

[38] Demir MM, Yilgor I, Yilgor E, Erman B. Electrospinning of polyurethane fibers. Polymer (Guildf). 2002;**43**:3303-3309

[39] Buchko CJ, Chen LC, Shen Y, Martin DC. Processing and microstructural characterization of porous biocompatible protein polymer thin films. Polymer (Guildf). 1999;**40**:7397-7407

[40] Zhang N et al. Pharmacokinetics, tissue distribution and anti-tumor effect of low density lipoprotein peptide conjugated submicron emulsions. Biomedicine & Pharmacotherapy. 2016;**82**:614-619

[41] Sill TJ, von Recum HA. Electrospinning: Applications in drug delivery and tissue engineering. Biomaterials. 2008;**29**:1989-2006

[42] Ki CS et al. Characterization of gelatin nanofiber prepared from gelatin–formic acid solution. Polymer (Guildf). 2005;**46**:5094-5102

[43] Han J, Chen TX, Branford-White CJ, Zhu LM. Electrospun shikonin-loaded PCL/PTMC composite fiber mats with potential biomedical applications. International Journal of Pharmaceutics. 2009;**667**:1-6

[44] Casper CL, Stephens JS, Tassi NG, Chase DB, Rabolt JF. Controlling surface morphology of electrospun polystyrene fibers: Effect of humidity and molecular weight in the electrospinning process. Macromolecules. 2004;**37**:573-578

[45] Bae H-S et al. Fabrication of highly porous PMMA electrospun fibers and their application in the removal of phenol and iodine. Journal of Polymer Research. 2013;**20**(158)

[46] Zeeb B, Saberi AH, Weiss J, McClements DJ. Formation and characterization of filled hydrogel beads based on calcium alginate: Factors influencing nanoemulsion retention and release. Food Hydrocolloids. 2015;**50**:27-36

[47] Jun Z, Hou H, Schaper A, Wendorff JH, Greiner A. Poly-L-lactide nanofibers by electrospinning—Influence of solution viscosity and electrical conductivity on fiber diameter and fiber morphology. E-Polymers. 2003;**3**:1-9

[48] Kim B, Park H, Lee S-H, Sigmund WM. Poly(acrylic acid) nanofibers by electrospinning. Materials Letters. 2005;**59**:829-832

[49] Geng X, Kwon O-H, Jang J. Electrospinning of chitosan dissolved in concentrated acetic acid solution. Biomaterials. 2005;**26**:5427-5432

[50] Geetha M, Asokamani R, Amit Kumar J, Ramalingam M. Nanofiber Composites for Biomedical Applications. 2017;**1**:117-175

[51] Pasricha R, Sachdev D. Nanofiber Composites for Biomedical Applications. 2017;**1**:157-196

[52] Polini A, Yang F. Nanofiber Composites for Biomedical Applications. 2017;**21**:622-629

[53] Naves LB, Almeida L, Rajamani L. Nanofiber Composites for Biomedical Applications. 2017. DOI: 10.1016/B978-0-08-100173-8.00011-9

[54] Chen H, Peng Y, Wu S, Tan LP. Electrospun 3D fibrous scaffolds for chronic wound repair. Materials (Basel). 2016;**9**:1-12

[55] Xu F, Li Luming CX. Fabrication of aligned side-by-side TiO_2/SnO_2 nanofibers via dual-opposite-spinneret electrospinning. Journal of Nanomaterials. 2012;**5**:1-5

[56] Nguyen TTT, Ghosh C, Hwang SG, Chanunpanich N, Park JS. Porous core/sheath composite nanofibers fabricated by coaxial electrospinning as a potential mat for drug release system. International Journal of Pharmaceutics. 2012;**439**:296-306

[57] Dhand C et al. Nanofiber Composites for Biomedical Applications. 2017;**1**:199-223

[58] Dai XY et al. Electrospun emodin polyvinylpyrrolidone blended nanofibrous membrane: A novel medicated biomaterial for drug delivery and accelerated wound healing. Journal of Materials Science. Materials in Medicine. 2012;**23**:2709-2716

[59] Goins A et al. Fabrication of core-shell nanofibers for controlled delivery of bromelain and salvianolic acid B for skin regeneration in wound therapeutics. Biomedical Materials. 2017;**12**:1-17

[60] Shan YH et al. Silk fibroin/gelatin electrospun nanofibrous dressing functionalized with astragaloside IV induces healing and anti-scar effects on burn wound. International Journal of Pharmaceutics. 2015;**479**:291-301

[61] Caserta S, Sabetta L, Simeone M, Guido S. Shear-induced coalescence in aqueous biopolymer mixtures. Chemical Engineering Science. 2005;**60**:1019-1027

[62] Haraway GD. The extracellular matrix in wound healing. Healthpoint. 2006;**800**:1-4

[63] Felgueiras HP, Amorim MTP. Functionalization of electrospun polymeric wound dressings with antimicrobial peptides. Colloids and Surfaces B: Biointerfaces. 2017;**156**:133-148

[64] Byung-Chul K et al. Fibroblasts from chronic wounds show altered TGF-β-signaling and decreased TGF-β Type II Receptor expression. Journal of Cellular Physiology. 2003;**195**:331-336

[65] Wysocki AB, Staiano-Coico L, Grinnell F. Wound fluid from chronic leg ulcers contains elevated levels of metalloproteinases MMP-2 and MMP-9. The Journal of Investigative Dermatology. 1993;**101**:64-68

[66] Veleirinho B et al. Nanofibrous poly(3-hydroxybutyrate-co-3-hydroxyvalerate)/chitosan scaffolds for skin regeneration. International Journal of Biological Macromolecules. 2012;**51**:343-350

[67] Rieger KA, Birch NP, Schiffman JD. Designing electrospun nanofiber mats to promote wound healing—A review. 2013;**1**:4531-4541

[68] Çinar Medeni Ö, Baltaci G, Doğan Vayvay G. Acute effect of kinesiotape muscle technique on hamstring flexibility and pain during stretching. Fiz Rehabilitation. 2015;**26**:73-77

[69] Dhivya S, Padma V, Santhini E. Wound dressings-a review. Biomedicine. 2015;**5**:24-28

[70] Gonyon DLJ, Zenn MR. Simple approach to the radiated scalp wound

using INTEGRA skin substitute. Annals of Plastic Surgery. 2003;**50**:315-320

[71] Jingwei X, Bing M, Lorwattanapongsa MP. Fabrication of novel 3D nanofiber scaffolds with anisotropic property and regular pores and their potential applications. Advanced Healthcare Materials. 2012;**1**:674-678

[72] Zhong SP, Zhang YZ, Lim CT. Tissue scaffolds for skin wound healing and dermal reconstruction. Wiley Interdisciplinary Reviews: Nanomedicine Nanobiotechnology. **2**:510-525

[73] Franz G. Polysaccharides in pharmacy: Current applications and futureconcepts. Planta Medica. 1989;**55**:493-497

[74] Medusheva EO et al. New medical materials with an integral lasting effect based on fibre-forming polymers. Fibre Chemistry. 2007;**39**:268-271

[75] Lim CT et al. Nanofiber technology: Current status and emerging developments. Progress in Polymer Science. 2017;**70**:1-17

[76] Liu Y et al. Effects of solution properties and electric field on the electrospinning of hyaluronic acid. Carbohydrate Polymers. 2011;**83**:1011-1015

[77] Hwa HK. Preparation and properties of electrospun poly(vinyl alcohol)/silver fiber web as wound dressings. Polymer Engineering and Science. 2006;**47**:43-49

[78] Gumusderelioglu M, Dalkiranoglu S, Aydin RST, Cakmak S. A novel dermal substitute based on biofunctionalized electrospun PCL nanofibrous matrix. Journal of Biomedical Materials Research. Part A. 2011;**98**:461-472

[79] Chen J-P, Chang G-Y, Chen J-K. Electrospun collagen/chitosan nanofibrous membrane as wound dressing. Colloids Surfaces A: Physicochemical and Engineering Aspects. 2008;**313-314**:183-188

[80] Tchemtchoua VT et al. Development of a chitosan nanofibrillar scaffold for skin repair and regeneration. Biomacromolecules. 2011;**12**:3194-3204

[81] Zhou Y et al. Electrospun water-soluble carboxyethyl chitosan/poly(vinyl alcohol) nanofibrous membrane as potential wound dressing for skin regeneration. Biomacromolecules. 2008;**9**:349-354

[82] Bonvallet PP et al. Microporous dermal-mimetic electrospun scaffolds pre-seeded with fibroblasts promote tissue regeneration in full-thickness skin wounds. PLoS One. 2015;**10**:1-17

[83] Suganya S et al. Aloe vera incorporated biomimetic nanofibrous scaffold: A regenerative approach for skin tissue engineering. Iranian Polymer Journal. 2014;**23**:237-248

[84] Gomes T et al. Polymeric nanoparticles of Brazilian red propolis extract: Preparation, characterization, antioxidant and leishmanicidal activity. Nanoscale Research Letters. 2016

[85] Sikareepaisan P, Suksamrarn A, Supaphol P. Electrospun gelatin fiber mats containing a herbal-Centella asiatica-extract and release characteristic of asiaticoside. Nanotechnology. 2008;**19**(15102)

[86] Yang Y et al. Electrospun fibers with plasmid bFGF polyplex loadings promote skin wound healing in diabetic rats. Molecular Pharmaceutics. 2011;**9**

[87] Rana D, Ratheesh G, Ramakrishna S, Ramalingam M. Nanofiber Composites for Biomedical Applications. 2017:**1**:325-344

[88] Kuppan P, Vasanthan KS, Sundaramurthi D, Krishnan UM, Sethuraman S. Development of poly(3-hydroxybutyrate-co-3-hydroxyvalerate) fibers for skin tissue engineering: Effects of topography, mechanical, and chemical stimuli. Biomacromolecules. 2011;**12**:3156-3165

[89] Kenawy ER et al. Release of tetracycline hydrochloride from electrospun poly(ethylene-co-vinylacetate), poly(lactic acid), and a blend. Journal of Controlled Release. 2002;**81**:57-64

[90] Kheradvar SA, Nourmohammadi J, Tabesh H, Bagheri B. Starch nanoparticle as a vitamin E-TPGS carrier loaded in silk fibroin-poly(vinyl alcohol)-Aloe vera nanofibrous dressing. Colloids Surfaces Biointerfaces. 2018;**166**:9-16

[91] Simões D et al. Recent advances on antimicrobial wound dressing: A review. European Journal of Pharmaceutics and Biopharmaceutics. 2018;**127**:130-141

[92] Zahedi P, Rezaeian I, Ranaei-Siadat SO, Jafari SH, Supaphol P. A review on wound dressings with an emphasis on electrospun nanofibrous polymeric bandages. Polymers for Advanced Technologies. 2010;**21**:77-95

[93] Rujitanaroj P, Pimpha N, Supaphol P. Wound-dressing materials with antibacterial activity from electrospun gelatin fiber mats containing silver nanoparticles. Polymer (Guildf). 2008;**49**:4723-4732

[94] Li C et al. Silver nanoparticle/chitosan oligosaccharide/poly(vinyl alcohol) nanofibers as wound dressings: A preclinical study. International Journal of Nanomedicine. 2013;**8**:4131

[95] Zhou JY, Zhang Z, Qian GS. Mesenchymal stem cells to treat diabetic neuropathy: A long and strenuous way from bench to the clinic. Cell Death Discovery. 2016;**2**:1-7

[96] Barani H. Antibacterial continuous nanofibrous hybrid yarn through in situ synthesis of silver nanoparticles: Preparation and characterization. Materials Science and Engineering: C. 2014;**43**:50-57

[97] Alberti TB. Desenvolvimento de nanomatrizes de álcool polivinílico e própolis com aplicação em cicatrização de pele. UFSC; 2016;**1**:1-63

Chapter 10

Nanofibers in Mucosal Drug and Vaccine Delivery

Josef Mašek, Eliška Mašková, Daniela Lubasová, Roman Špánek, Milan Raška and Jaroslav Turánek

Abstract

Successful mucosal administration and delivery of drugs still pose a great challenge. However, the possibility to deliver not only small drug molecules but also macromolecular drugs and nanoparticles via mucosal surfaces represents a great opportunity. Rapid onset of drug action, avoidance of first-pass metabolism, and high immunocompetence of mucosa are some of the important features for mucosal drug and vaccine delivery. The use of mucoadhesive drug delivery systems, systems with fast dissolving properties, and nanomaterials with mucus penetration properties are examples of successful strategies to achieve effective mucosal drug and vaccine delivery. Non-keratinized mucosa of the oral cavity, the nasal and vaginal mucosa represent favorable sites of drug administration. Polymer nanofibers have attracted much attention because of remarkable characteristics such as a large surface area to volume ratio and high porosity. Nanofibers have been extensively used for different biomedical applications including wound dressing, tissue engineering, and drug delivery. Among their fabrication methods, the introduction of electrospinning technique was an important step toward achieving the goal of large scale industrial production of nanofiber-based drug delivery systems used in mucosal applications. This chapter provides an overview on all aspects of mucosal drug and vaccine delivery using nanofibers.

Keywords: nanofibers, electrospinning, transmucosal drug delivery, mucoadhesive formulation, mucus penetration nanoparticles, first-pass metabolism, mucosal vaccines

1. Introduction

Mucosal drug delivery is an alternative method of systemic drug delivery that offers numerous benefits over/parenteral and oral administration. Mucosal surfaces, particularly oral, nasal, and vaginal, have been widely explored for systemic delivery of drugs. Drugs that are absorbed via mucosal surfaces directly enter the systemic circulation and bypass the gastrointestinal tract including first-pass metabolism in the liver. Rapid onset of drug action is another advantage of oral, nasal, and vaginal mucosae drug administration. A lot of efforts have been devoted to the discovery of efficient delivery of vaccine antigens to mucosal sites that enhance uptake by local antigen-presenting cells in order to generate protective mucosal immune responses. Potent mucosal adjuvants and suitable mucosal vaccine delivery systems are crucial steps on the way to effective mucosal vaccines.

On the other hand, several challenges need to be overcome to successfully deliver drug molecules. Poor water solubility of numerous drugs, macromolecular nature of newly developed biologically active molecules, including therapeutic peptides, proteins, and nucleic acids, are some examples of challenging features that need to be overcome by developing new delivery approaches, devices, and dosage forms. Mucoadhesive drug formulations, rapidly disintegrating formulations, and formulations with mucus penetration properties enabling mucosal delivery of such therapeutic molecules have made great progress over the last decade.

Anatomical and physiological functions of different types of mucosa need to be taken into consideration when formulating new drug delivery systems. These need to respect the flexibility of mucosal surfaces, the flow of body fluids such as saliva or vaginal fluid, ciliary movement on the nasal mucosa, presence of mucosal absorption barriers, including the mucus layer and keratinization of some surfaces, etc.

Nanofibers represent an interesting opportunity to tackle some of the difficulties in mucosal drug delivery. Nanofibers represent an almost universal platform the properties of which can be tailored according to specific demands in terms of their composition as well as surface modifications. Advanced features of nanofibers include possible mucoadhesive properties, fast disintegration, controlled drug release, formulation of small drugs of different nature, and formulation of therapeutic macromolecules. These properties can help solve problems with mucosal delivery of poorly water-soluble drugs, drugs with fast liver metabolism, therapeutic proteins, and antigens for mucosal immunization. Taken together, formulations based on nanofibers intended for mucosal applications represent a new trend in drug delivery.

2. Histology and barrier properties of mucosal surfaces

Histology, physiology, and barrier functions of mucosal surfaces play a critical role in mucosal drug and vaccine delivery. All aspects have to be taken into consideration when designing a mucosal drug delivery system. Also, appropriate drug candidates for mucosal delivery have to be selected, not only with respect to physical-chemical properties of drug molecule, but also anatomical and physiological aspects of the intended site of administration and their deep knowledge are essential for successful delivery of drugs and vaccines.

2.1 Histology of mucosal surfaces

2.1.1 Oral mucosa

The mucosa of oral cavity is divided into the buccal, sublingual, gingival, palatal, and labial regions. The mucosa of each region is of specific histological and functional characteristics. Oral mucosa consists of three layers: a stratified squamous epithelium, composed of several cell layers, below which lies the basement membrane, and finally the connective tissue divided into the lamina propria and submucosa, which comprise a number of vascular capillaries [1]. Drugs absorbed via the oromucosal route of administration are absorbed through these capillaries and gain access to the systemic circulation [2, 3].

Three major types of epithelium located in different regions of the oral cavity differ in the degree of keratinization, namely masticatory, specialized, and lining mucosa. The masticatory epithelium is keratinized (100–200-μm thick) and covers the gingival region and the hard palate. The specialized epithelium is stratified, keratinized, and covers the dorsal surface of the tongue. The lining mucosa covers buccal and sublingual regions of the oral cavity. The epithelial layer of the buccal and sublingual mucosa is non-keratinized, with variation in thickness [4]. The lining mucosa exhibits high permeability for different drugs, and thus is an interesting site for drug administration.

The oral epithelium is covered by a 70–100-μm thick film of saliva, the secretion from salivary glands. The daily production of saliva secreted into the oral cavity is between 0.5 and 2 L. Continuous production of saliva significantly impacts drug residence time after administration within the oral cavity, a phenomenon known as saliva washout [5].

Mucus is the intercellular ground matrix secreted by the sublingual and salivary glands, which is bound to the apical cell surface and acts as a protective layer for the cells below [6]. It is also a viscoelastic hydrogel consisting of the water-insoluble glycoproteins, water, and small quantities of different proteins, enzymes, electrolytes, and nucleic acids. The mucus layer carries a negative charge due to a high content of sialic acid and forms a strongly cohesive gel structure that binds to the epithelial cells.

The mucus layer varies in thickness from 40 to 300 μm. The mucus layer plays a critical role in the function of different mucoadhesive drug delivery systems which work on the principle of mucoadhesion, and thus prolong the dosage form retention time at the site of administration [7].

2.1.2 Nasal mucosa

The total surface area available in the human nasal mucosa is estimated to be about 180 cm^2. Most of the surface is covered with highly vascularized respiratory mucosa and only a small surface area of the nasal cavity is covered by olfactory mucosa. Due to the presence of microvilli on the apical cell surface, the effective surface area for drug absorption is relatively high [8]. The nasal vestibule is lined by stratified squamous epithelium which gradually transitions in the valve region with a ciliated, pseudostratified, columnar epithelium characterized by presence of mucus-secreting goblet cells. Mucus lies over the epithelium as a protective layer and the mucociliary apparatus filters the air. Mucus is secreted at a flow rate of 5 mm/min and this fast renewal rate means that particles are eliminated from the nasal cavity in less than 20 min. The nasal mucosa is about 2–4-mm thick and composed of two distinct layers: periciliary layer and superficial layer [5, 9].

Enzymes and peptidases contained in the mucus, its constant secretion, and nasal clearance mechanisms significantly reduce the ability of drugs to penetrate through the epithelium [10]. Studies have shown that rapid systemic delivery of topically applied drugs can be achieved after intranasal administration due to the permeable properties of the respiratory epithelia and highly vascularized nature of the adjacent submucosa.

Immunologically specialized region localized at the gateway of the respiratory and alimentary tract consists of the palatine tonsils, nasopharyngeal tonsils (adenoid), lingual tonsils, and tubal tonsils and is designated as Waldeyer's ring or nasal-associated lymphoid tissue (NALT). It represents a site of intimate contact between exogenous antigens and the host aerodigestive tract. With exception of adenoids and tubal tonsils covered by pseudostratified ciliated columnar epithelium (respiratory epithelium) the palatine and lingual tonsil are covered by stratified non-keratinized squamous epithelium with many invaginations allowing the enhanced exposure of foreign antigens to the underlying cryptal lymphatic tissue. Therefore, nasal mucosa is one of intensively explored sites for vaccine administration.

2.1.3 Vaginal mucosa

The vaginal mucosa is composed of four histological layers: the stratified squamous epithelium (with underlying basement membrane), the elastic lamina propria (a dense connective tissue layer which projects papillae into the overlying epithelium), the fibromuscular layer comprising two layers of smooth muscle, and the tunica adventitia, consisting of areolar connective tissue [5].

The vaginal epithelium is composed of non-cornified, stratified squamous cells. Several layers of cuboidal cells are close to the basement membrane and become flattened as they move toward the luminal surface. The thickness of vaginal epithelium is around 200–300 μm, and is influenced by physiological factors, e.g., variability in estrogen concentration [11].

Vaginal surface is coated with a layer of cervicovaginal fluid. The fluid contains secretions from the cervix in the form of mucins and secretions from the Bartholin's and Skene's glands, endometrial fluid, and fluid transuded from the vascular bed of the vaginal tissue. It also contains a large number of squamous epithelial cells, enzymes, proteins, carbohydrates, and amino acids. The acidic vaginal environment is maintained by production of lactic acid by lactobacilli but the exact pH value is influenced by the presence of cervical mucus, the amount of vaginal fluid, infections, and other factors. Cervical mucus forms a semipermeable viscoelastic barrier at the vaginal surface. High constant production of cervicovaginal fluid influences the properties of different drug delivery systems, the residence time after administration as well as the rate of drug release and absorption [12].

2.2 Barrier functions of mucosal surfaces

2.2.1 Oral mucosa

The rate of drug absorption following oromucosal administration is influenced by the permeability of the buccal and sublingual mucosa, physical-chemical properties of the delivered drug, and other factors, namely the presence and properties of mucus, saliva production, movement of the oral tissues during speaking, food and drink intake, etc. The mucus layer is the main natural barrier of mucosa against penetration of different pathogens and foreign particles. One the other hand, mucus layer is one of the main absorption barriers to a variety of drugs, including nanoparticle-based drugs and vaccine formulations [5].

Drug permeability through the oral cavity mucosa represents a major limiting factor in transmucosal drug delivery. Mechanically stressed areas are keratinized and impermeable to water, which makes such areas unfavorable for drug delivery. On the other hand, more permeable non-keratinized buccal and sublingual epithelia make such regions of the oral cavity attractive sites for drug delivery and a great number of active ingredients are currently being explored in terms of transmucosal drug delivery [13].

One of the main permeability barriers of oral mucosa is represented by the presence of a layer of extracellular lipidic material coming from membrane-coating granules (MCGs). They are present in both keratinized and non-keratinized epithelia, but the composition of lipidic material is different between the two types [14].

Passive diffusion of drugs and drug carriers through mucosal surfaces is generally considered the primary mechanism responsible for the transport of drugs across the oral mucosa [15]. However, active transport mechanisms are connected to the activity of different types of immune cells widely present in oral mucosal tissues, especially in sublingual and buccal regions. Such immune cells, mainly different types of dendritic cells, act as an immunological barrier between the body and foreign stimuli (including pathogens and potential allergens), mainly coming to the body with food and drink. These cells are responsible for tolerogenic, or opposite reaction to such stimuli.

Another barrier to drug permeability across oromucosal surfaces is enzymatic degradation. Saliva contains some enzymes that are able to metabolize some peptide and protein therapeutics. This fact leads to reduced bioavailability of protein-based therapeutics after oromucosal administration.

The flow of saliva makes it difficult for a drug delivery system to retain the released drug at the site of administration and, therefore, mucoadhesive drug formulations are the preferred dosage form for oromucosal drug delivery [5].

2.2.2 Nasal mucosa

The main barrier functions of nasal mucosa include the role of the mucus layer and the process of mucociliary clearance. For systemically acting drugs, the mucus layer acts as a diffusion barrier. The rate of the diffusion through the mucus layer is dependent on several factors involving thickness and viscosity of the mucus, and drug physical-chemical properties. The thickness of the layer of mucus in the nasal cavity is a few microns and, therefore, in contrast to other mucosal surfaces, it does not represent a substantial diffusion barrier. The rate of mucociliary clearance influences the contact time of epithelia and a drug delivery formulation and also the drug molecules themselves [9].

Transport across epithelial barriers with respect to the nasal route, and drug permeation via the paracellular route are limited because of the presence of tight junctions. Therefore, passive diffusion via the transcellular route is the main transport pathway across the nasal epithelium [16].

Physical-chemical factors that influence nasal drug absorption are the molecular weight and lipophilicity. Polar and ionized molecules show poor permeability. On the other hand, drugs with systemic action currently administered by the intranasal route are of low molecular weight and lipophilic nature. Moreover, some therapeutic peptides, e.g., salmon calcitonin and oxytocin, also exhibit some level of systemic nasal absorption [5].

2.2.3 Vaginal mucosa

Properties of the vaginal epithelium such as thickness and barrier functions are depending on several factors, including age and hormonal activity [12]. The above mentioned factors affect both vaginal fluid production and the amount of enzymes contained in the fluid. The surface of vaginal epithelium contains a variety of enzymes which can metabolize the delivered drugs, especially therapeutic peptides and proteins. The present proteases are the main barrier for absorption of such drugs into the blood as they are degraded before they cross the epithelium and can reach systemic circulation. Cervicovaginal fluid acts as a dissolution medium that enables transfer of drugs from the dosage form into the tissue. It has also been claimed that drugs can directly be transferred from dosage form to epithelial tissues. This means that a portion of the released drug molecules can overcome the fluid compartment. The pH in the vagina is around 4, thus strongly acidic under normal conditions. These conditions can lead to rapid drug degradation.

Systemic mucosal delivery through vaginal epithelium has several advantages, including ease of administration, avoidance of drug liver first-pass metabolism, relatively low metabolic activity, high permeability, prolonged retention, and the potential for sustained release from drug delivery formulations with controlled release properties. Important physical-chemical properties of drugs delivered through vaginal mucosa include drug solubility in the vaginal fluid, tissue permeability, chemical stability of drugs under certain pH conditions, and drug lipophilicity [5].

2.3 Mucosal vaccines

Mucosal surfaces, such as those of the respiratory, gastrointestinal, and genital tracts, act as the first line of defense against environmental pathogens. Although immunization by mucosally applied vaccines has had a long history with numerous micro-organisms and a variety of administration routes, only a few such vaccines are currently used in human medicine. Nevertheless, these examples (e.g., poliomyelitis, influenza, cholera, *Salmonella typhi*) clearly demonstrate the validity of this strategy. There are many reasons for the development of mucosal vaccines as an attractive alternative to those administered by systemic routes. Most importantly, protective immune responses can be induced at the relevant mucosal sites of pathogen entry by mucosal delivery of vaccines, and thus the enormous potential of immunity in mucosal tissues and their associated secretory glands remains to be exploited in vaccinology. Better understanding of mucosal immune system together with new nanomaterials, including nanofibers and engineering of recombinant antigens, enhanced a long-lasting effort to develop mucosal vaccines capable of effectively inducing both mucosal and systemic immune responses, thereby resulting in two layers of host protection.

Mucosal vaccination, in contrast to other routes of vaccine administration, is of particular interest since it can enhance immune responses, mainly secretory IgA, which defends the portal of entry of various infectious pathogens [17]. Since the route of vaccine administration has a significant effect on the resultant immune response, much effort has been made to explore novel mucosal vaccine delivery routes, briefly described in this chapter.

The noninvasive needle-free vaccine delivery mode (nonparenteral routes of application) has become a global priority, both to eliminate the risk of improper and unsafe needle use and to simplify vaccination procedures. Development of alternative vaccine delivery methods, including mucosal routes, becomes a prominent field of vaccine research and represents a challenge for new biocompatible materials, especially nanomaterials. The vaccine administration route significantly affects immune responses regarding the intensity and quality (class-specified Ig, TH1/TH2 balance, anergy).

Furthermore, for vaccination campaigns organized to stop epidemics of mucosally transmitted infections, mass immunization by mucosal routes is likely to be more practicable and less expensive than immunization by systemic routes. In addition, many parents already hesitate to subject their young children to repeated needle sticks. The factors like reduced cost of mucosal vaccines must be taken into consideration. Economy of mucosal vaccines is based on both production/distribution and application levels. The purity of mucosally delivered vaccines, including endotoxin contamination, is less critical than for injectable vaccines. Finally, mucosal vaccine delivery does not require sterile syringes and needles or personnel trained in their use and disposal, although spray devices or other applicators may be needed for intranasal and other routes of administration. From this point of view, sublingual application of vaccines represents the easiest and the most favorable modes.

Nevertheless, there are some drawbacks in terms of mucosal vaccination, particularly the uncertainty of the amount of effectively delivered antigen following its mucosal administration. Another problem is the limited uptake of intact protein and polysaccharide antigens at mucosal surfaces. Moreover, potential for degradation of antigens, especially in the gastrointestinal tract following oral administration is another issue. Therefore, as compared to oromucosal administration, significantly higher doses of vaccine antigens must be administered orally to induce measurable immune responses. To overcome such difficulties, various particulate antigen delivery systems have been designed.

Successful development of future mucosal vaccines is based on three basic pillars critical for inducing the effective immune response. These pillars are adjuvants, application/delivery systems, and antigens.

Adjuvants represent one of the key issues in all vaccine development. The identification and development of appropriate mucosal adjuvants to enhance the desired aspects of the immune response against the antigens represent a special challenge because adjuvants for mucosal application involve requirements different from those for parenteral use. On the other hand, this affords greater opportunities for discovery by exploiting the growing understanding of the mechanisms whereby mucosal pathogens and commensals interact with the immune system. Cholera toxin (CT) and related heat-labile enterotoxins such as *Escherichia coli* labile toxin (LT) are classic examples of mucosal adjuvants for their ability to break immune tolerance [18]. A major concern for human application has been to separate the toxicity of these molecules from their adjuvant properties. Different approaches were used based on targeted mutations in the enzymatically active site of the A subunit, the selection of various types of toxins including mutants that have different ganglioside (receptor)-binding B subunits and the use of nontoxic B subunits alone or coupled to vaccine antigens. Intranasal administration has generated interest and concern with the finding that CT and LT can undergo retrograde migration along neurons, potentially reaching the brain via the olfactory nerve, and thus leading to neurological pathology. An intranasal influenza vaccine which contained a low dose of LT as an adjuvant was introduced in Switzerland, but it was withdrawn due to suspicion of causing Bell's palsy in some recipients [19]. Facial nerve palsy was another side effect demonstrated after intranasal application of a flu vaccine. As an alternative route, sublingual vaccination has been demonstrated to achieve similar immune responses but without the risk of retrograde transmission to the brain [20].

Other molecules derived from microorganisms and falling into the category of pathogen associated molecular pattern (PAMP) have also been shown to have immunomodulatory properties. Ligand for Toll-like receptors, NOD1/2 receptors, and inductors of inflammasomes are examples of such molecules (e.g., monophosphoryl lipid A, "CpG" oligodeoxynucleotides, muramyl glycopeptides, flagellin). Nevertheless, none of these has been investigated as extensively as the heat-labile enterotoxins.

Delivery systems for mucosal immunization are functionally related to adjuvants for mucosal vaccines as well as to antigen formulations. The principal task for delivery systems is to keep antigen on mucosal surface and to enhance its penetration into submucosal tissue to be accessible in sufficient amount to immune cells, especially to dendritic cells. Various gels and films are available for buccal application of drugs, while sprays or droplets are used for intranasal application of influenza vaccines. Oral application of vaccines is achieved mainly by various capsules containing antigens or simply by administration of the vaccine on a spoon (polio).

While some formulations for nasal and oral delivery of vaccines are available, suitable delivery systems for sublingual vaccination are at their door step of development. In this chapter, we describe the mucoadhesive nanofiber-based film fulfilling the role of delivery system for sublingual immunization. This system is compatible with antigens formulated as proteoliposomes and immunostimulating complexes (ISCOMs), virus and virus-like particles, bacteria and bacterial ghosts, plasmid DNA, polymeric nanoparticles, and a simple free antigen. Such formulations also allow the sustained release of vaccine antigens and provide some protection against removal from the site of application by saliva flow and tongue movement. Also, enhancers of mucosal penetration can be combined with antigens and mucosal adjuvants, which make the mucoadhesive nanofiber-based films a universal platform for mucosal sublingual vaccination. The important aspect of the mucoadhesive nanofiber-based film is its suitability for industrial production as the crucial factor of pharmacoeconomy [21].

In conclusion, sublingual immunization represents the route of a big hidden potential for development of effective, cheap, and safe vaccines for prevention of infectious diseases as well as the treatment of allergy.

2.4 Penetration enhancers

Penetration enhancers, sometimes also referred to as chemical enhancers, are functional pharmaceutical excipients that possess the ability to modify the physical-chemical properties of mucosal barriers. The reason for co-administration of penetration enhancers is to enhance penetration of drugs through different barriers of mucosa including mucus layer, extracellular lipid layer originating from MCGs, enzymatic barrier, and others. The target compartment of the delivered drug can differ according to the mode of its action, e.g., blood circulation in systemic drug delivery, specialized immune cells present in epithelium or submucosal tissue in vaccine delivery and local cell therapy, etc. Penetration enhancers can potentially facilitate the systemic absorption of a wide range of drugs, including large therapeutic molecules such as polypeptides, proteins, nucleic acids [22], and therapeutic nanoparticles.

Different surfactants, bile salts (deoxycholic, ursocholic, and taurocholic acids), ethylenediaminetetraacetic acid (EDTA), fatty acids, and amino acids are the most frequently explored chemicals used in mucosal enhancing of drug permeability, and thus its bioavailability [23, 24]. An important concern related to penetration enhancers is the capacity of adjacent tissues to tolerate the effects of penetration enhancers.

Penetration enhancers act usually by a combination of different mechanisms. Some penetration enhancers are amphipathic in nature, and thus associate and influence bilayers of the cell membranes, increase the fluidity and permeability of membranes and finally promote transcellular transport. Penetration enhancers can interact with tight junctions between epithelial cells that cause facilitation of paracellular drug transport. Another group of penetration enhancers, known as mucolytic agents (e.g., acetylcysteine) influence the integrity of the mucus layer.

Formulation of drugs into nanofibers, especially nanofibers with mucoadhesive properties, enhances drug absorption via mucosal surfaces in general. Moreover, combination of nanofiber-based formulations and chemical enhancers can increase penetration of drug molecules, or nanoparticle-based drug and vaccine delivery systems. Nanofibers themselves have extraordinary capacity to combine different types of chemicals including combination of drugs and penetration enhancers, whether using electrospinning technique for nanofiber production, or another technique. As an example, mucosal penetration of model mucus penetration poly(lactic-co-glycolic) acid (PLGA) nanoparticles from nanofiber-based mucoadhesive film into porcine sublingual and buccal mucosa was enhanced using sodium deoxycholate as penetration enhancer [21].

3. Transmucosal delivery of drugs and vaccines

Different parts of the human body are covered by mucosa with different features and barrier properties for drug delivery. Some areas are more accessible than others. In principle, there are two ways of mucosal drug delivery—local delivery, e.g., antimicrobials, anti-inflammatory drugs, etc., and systemic drug delivery. Whereas many indications for drugs intended for local treatment exist and include all mucosal surfaces, systemic drug delivery is a completely different task and only a few parts of mucosa are suitable sites for systemic drug administration and absorption. Moreover, only several drugs with suitable physical-chemical properties have been administered via transmucosal delivery. On the other hand, mucosal drug absorption has enormous potential, and different strategies including penetration enhancers, nanoparticle-based drugs, and mucoadhesive drug formulations have been employed.

Because of easy accessibility, high vascularization and a relatively low thickness, oral mucosa, especially its sublingual and buccal parts, is a preferred site of systemic drug administration. Transmucosal delivery of certain drugs can result in a rapid onset of their action, thus having the potential to replace the injection administration of some specific drug molecules [25]. Oromucosal delivery of nitroglycerin brings many benefits for patients, and today, nitroglycerin is one of the most frequent drug molecules delivered via oral mucosa. Exploration of cardiovascular drugs (e.g., captopril, verapamil) has shown promising results. Oral transmucosal delivery of analgesics has attracted substantial attention. As an example, oromucosally administered fentanyl is designed for rapid noninvasive delivery of analgesia for severe pain treatment.

Oromucosal delivery of sedatives and hypnotics has shown favorable results with clinical advantages over other routes of administration. Another example is drugs for erectile dysfunction and transmucosal formulations of hormones, e.g., testosterone and estrogen. Transmucosal spray formulation of insulin (Oral-lyn®, Generex) is a great example of the potential of oral mucosa for systemic macromolecular drug delivery.

Transmucosal nasal drug delivery is another interesting site for systemic drug delivery. The considerable blood supply of nasal mucosa provides efficient systemic drug absorption and enables direct access to the systemic circulation for drugs. In nasal drug delivery, limited nasal capacity often results in partial swallowing of the instilled drug as the instilled volumes exceed the limited capacity of the nasal mucosa. Therefore, the administered dose of the drug is partially swallowed, and drug absorption is in part transmucosal, in part gastrointestinal. Formulation of drugs into different mucoadhesive dosage forms and nasal inserts can be beneficial in this regard. Several drugs have been successfully administered including active ingredients with hormonal activity (salmon calcitonin, oxytocin, desmopressin) and small drugs (e.g., sumatriptan, zolmitriptan, dihydroergotamine).

As nasal mucosa is the first barrier which must be conquered by pathogens, nasal mucosa is very immunocompetent. Many studies have shown that even small amounts of antigen can elicit a strong protective response. Intranasal administration, similarly to oromucosal, seems to be the best strategy for barrier vaccinations following the outbreak of highly infectious diseases, because less erudite persons (e.g., nurses) can provide mass vaccination. FluMist® (US) and Fluenz® (Europe) are examples of live attenuated influenza vaccines.

The vaginal mucosa offers many advantages as a site for drug delivery, including easy access, prolonged residence time interval of drug availability, avoidance of first-pass metabolism, and relatively low activity of proteases and other enzymes. The vagina has a rich vascularization and a large surface area due to the folds in the mucosa (rugae) making it ideal for high absorption of drug molecules. Administration of drugs through vaginal mucosa represents an interesting alternative to oral administration for drugs treating osteoporosis, hormone replacement therapy, contraception, infections, and others. Mucoadhesive polymers are often used in formulations for vaginal drug delivery systems to prolong retention time of drug delivery systems [26].

Mucosal immune responses in the genital tract can be induced by the administration of antigen to mucosal surfaces. IgA antibodies in the vaginal tract are essential as a first line defense against pathogens that enter the body through vaginal mucosa. Immune responses of various types of vaginally administered vaccines have been investigated. Both vaginal and serum IgA and IgG levels have been enhanced following vaginal vaccine administration [27].

The nasal mucosa vaccination induces preferentially mucosa-associated immune responses. There are several vaccination approaches to induce both mucosal and systemic immune responses (antibodies and cell-mediated immunity), for example by heterologous immunization by systemic followed by mucosal routes. Combining of parenteral and mucosal administration of antigen is required because parenteral vaccines are notoriously inefficient for stimulating immune responses in mucosal tissues; and on the other hand, mucosal vaccination, particularly that administered by oral route to subjects without antecedent contact with vaccination antigen leads to induction of the "oral tolerance" phenomenon consisting in induction of IgA-mediated immune responses in mucosal compartments but dominantly cell-mediated antigen-specific tolerance in systemic compartment. In contrast, sublingual (SL) vaccination represents a route effectively stimulating both systemic and mucosal antibody- and cell-mediated immune responses. Sublingual mucosa is the place of vaccine administration which in contrast to other mucosally effective intranasal routes does not elicit neurotoxic effects as demonstrated for example with inactivated influenza virus administered SL together with a mucosal adjuvant which did not migrate to or replicate in the nerve system [28]. Furthermore, sublingual mucosa may be useful as a delivery site for mucosal vaccines because the sublingual epithelium harbors a dense lattice of dendritic cells (DC), and that using mucosal adjuvants mobilizes DCs within the sublingual epithelium. These cells migrate to the above mentioned proximal draining lymph nodes (submaxillary and superficial cervical lymph nodes), on uptake of the sublingual vaccine antigens. It is important that sublingual immunization induces antigen-specific immune responses in the female reproductive tract in addition to the respiratory tract and oral/nasal cavity [29]. Another mucosal administration route, orogastric route, can induce strong mucosal responses, especially secretory IgA in the small intestine, proximal colon, and mammary and salivary glands but it is poorly efficient for disseminating these responses to the distal segments of the gut and to the respiratory and reproductive tracts. Moreover, orogastric immunization requires substantially more antigen application because of intensive degradation. In addition to orogastric, nasal, and sublingual vaccines, transcutaneous immunizations are now part of a new generation of mucosal vaccines [30].

4. Nanofibers in mucosal drug delivery applications

4.1 Mucoadhesion and mucoadhesive properties of nanofibers

Mucoadhesion is defined as adhesion between a mucosal surface and a surface of another material. Mucoadhesive dosage forms have recently attracted much attention from pharmaceutical research as well as from pharmaceutical industry due to substantial improvements in mucosal drug delivery. Increasing the residence time of drug formulations at the site of administration automatically leads to much more effective transmucosal drug delivery, drug bioavailability, and results in increased therapeutic efficiency, thus lowering the drug dose needed [13]. Instead of small drug molecules, mucoadhesive formulations enable delivery of therapeutic biologicals such as peptides, proteins, antibodies, and nucleic acids through a variety of routes of administration such as oromucosal, ocular, nasal, and vaginal. Moreover, mucosal delivery of nanoparticle-based therapeutic formulations can be achieved by materials with mucoadhesive properties, mucus penetration formulations, and their combinations. Taste masking properties are of importance for mucoadhesive formulations intended for oromucosal administration.

Nanofibers made up of mucoadhesive polymers exhibit one of new trends in mucosal drug delivery. Due to their extremely large surface area, unique surface topology, and porosity, nanofibers are known to significantly improve the adhesiveness of the mucoadhesive drug delivery systems utilizing nanofibers for their construction. Architecture of nanofibers significantly intensifies the intimate contact between the nanofiber-based products and mucosal surface, and high drug concentration at the site of administration is achieved (**Figure 1**). Moreover, their ability to enhance drug solubility makes nanofibers an almost ideal platform for transmucosal drug delivery.

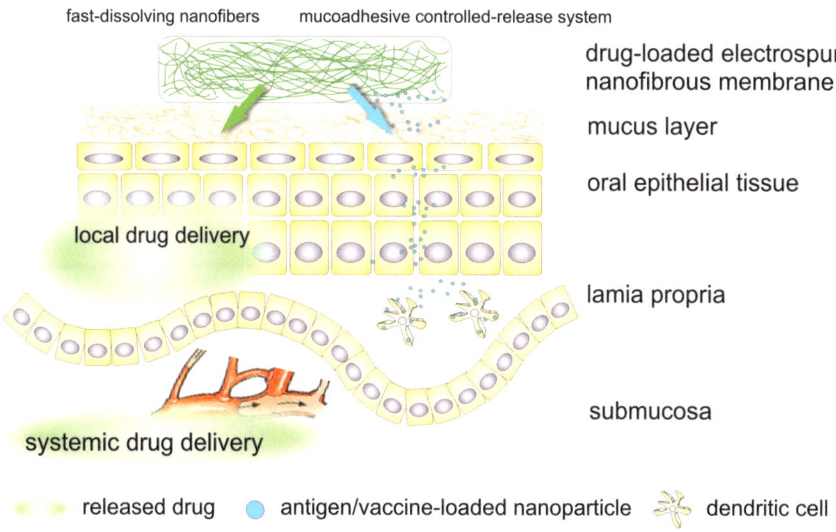

Figure 1.
Schematic representation of drug and vaccine delivery after mucosal administration of nanofiber-based drug delivery system.

The use of mucoadhesive polymers is conditioned by their ability to form nanofibers. A variety of factors affect the ability of polymers to form nanofibers as well as the mucoadhesive properties of polymers, including molecular weight, chain flexibility, hydrogen bonding capacity, cross-linking density, charge and degree of ionization of a polymer, concentration, and hydration (swelling) of a polymer [3].

These are reasons why only a few mucoadhesive polymers have been tested for nanofiber-based drug delivery system formulations. Chitosan (cationic polymer), hyaluronic acid, sodium alginate, sodium carboxymehylcellulose (anionic polymers), different cellulose derivatives, poly(ethylene oxide), and polyvinylpyrrolidone (nonionic polymers) are excellent examples of conventional mucoadhesive materials utilized for construction of nanofiber-based mucoadhesive drug delivery systems. A group of thiolated polymers, e.g., thiolated chitosan, are representatives of next-generation mucoadhesive materials [3, 13]. Mucoadhesive nanofibrous membrane made of chitosan/PEO is visualized in **Figure 2** as an example.

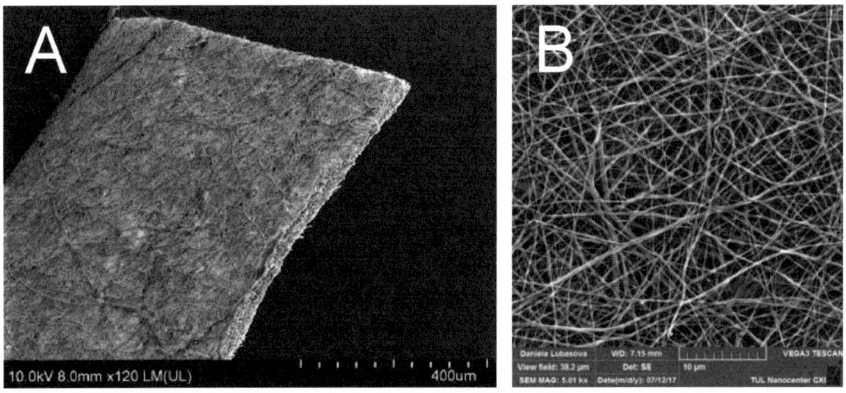

Figure 2.
Electrospun nanofibrous membrane fabricated using Nanospider technology (A) and detail of mucoadhesive chitosan/PEO nanofibers (B).

4.1.1 Chitosan

Chitosan is a linear polysaccharide composed of randomly distributed D-glucosamine and N-acetyl-D-glucosamine obtained by deacetylation of chitin (**Figure 3**). Because of the broad chemistry of chitosan, which covers different degrees of deacetylation, a range of molecular weights and different distribution of the acetyl groups along the polymeric chain, chitosans can provide a number of physical-chemical as well as biological properties. Mucoadhesion of chitosan occurs due to the electrostatic interactions of amino groups of chitosan and the sialic groups of mucin in the mucus layer.

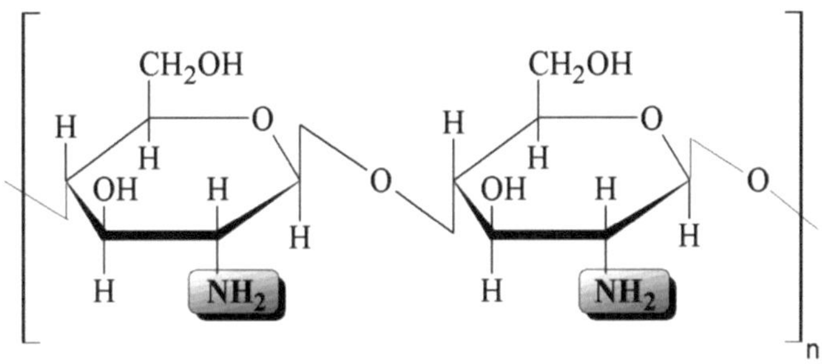

Figure 3.
Structural formula of chitosan.

Several studies have explored mucoadhesive properties of chitosan-based nanofibers. As an example, Lancina et al. have produced chitosan-based nanofiber mats capable of delivering insulin via the buccal mucosa. Chitosan was electrospun into nanofibers using poly(ethylene oxide) (PEO) as a carrier molecule. Insulin release rates were determined and showed no reduction in bioactivity due to electrospinning. Buccal permeation of insulin was significantly facilitated as compared to free insulin. Taken together, this work demonstrates that chitosan-based nanofibers have the potential to serve as a transbuccal insulin delivery vehicle [31]. In another study, mucoadhesive fibers of zein/chitosan have been prepared by electrospinning to study the encapsulation efficiency and release of tocopherol. The addition of the acidic chitosan solution to the zein containing tocopherol has improved the mucoadhesive properties of the final composite nanofibers [32]. Mucoadhesive hybrid electrospun chitosan/phospholipid nanofibers intended for drug-delivery applications were produced by Mendes et al. [33]. Nanofibrous membranes intended for local delivery of an antimicrobial agent in combination with poly(hexamethylene biguanide) hydrochloride were produced by electrospinning of chitosan/PEO solution. Inhibition of bacterial growth for both *Escherichia coli* and *Staphylococcus aureus* were achieved using nanofiberous membranes [34].

4.1.2 Cellulose derivatives

Cellulose (**Figure 4**) mucoadhesive derivatives are a wide group of pharmaceutical excipients and cover both nonionic polymers, including hydroxypropylmethylcellulose (HPMC), hydroxyethylcellulose (HEC), hydroxypropylcellulose (HPC), methylcellulose (MC), and anionic derivatives, e.g., carboxymethylcellulose (CMC).

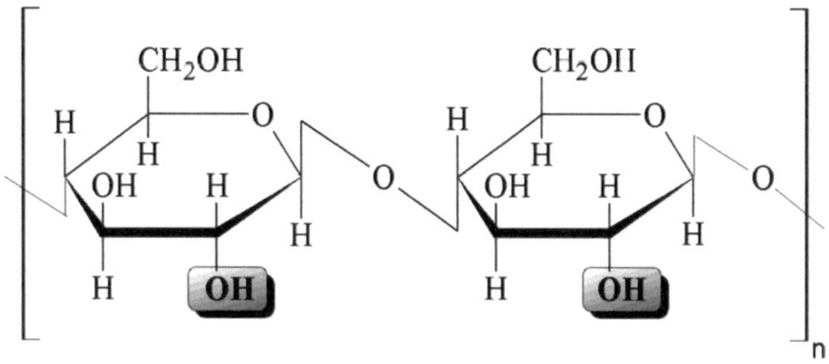

Figure 4.
Structural formula of cellulose.

The aim of the study exploring carboxymethylcellulose as a mucoadhesive agent for nanofiber formation was to develop a progesterone-loaded mucoadhesive system for vaginal application with sustained release. Presently, two dosage form options are being considered: direct compression of nanofibers into tablets for vaginal insertion and winding bundles of the fiber into a miniature tampon [35, 36].

In another work, nanofiber-based indomethacin films were prepared using different grades of methylcellulose, polyvinylpyrrolidone, and Tween® 80 by electrospinning. The addition of Tween® 80 to polyvinylpyrrolidone formulations significantly improved their wettability. Moreover, nanofiber-based patches containing methylcellulose and Tween® 80 were found to exhibit the highest permeation of indomethacin across porcine mucosa without significantly affecting the ultrastructure of the oral mucosa [37].

El-Newehy et al. demonstrated the preparation of HPC-based nanofibers. They found that the thermal stability and mechanical properties of nanofiber mats were dramatically enhanced with the addition of HPC to polyvinyl acetate or polyvinylpyrrolidone. The in vitro sustained release of an incorporated model drug, diclofenac sodium, was controlled when loaded into electrospun nanofibers of HPC with either PVA or PVP [38].

4.1.3 Poly(ethylene oxide)

Mucoadhesive glutamine-loaded poly(ethylene oxide) (**Figure 5**) electrospun nanofibers were prepared by Tort et al. The effect of different polyelectrolytes on resultant properties of nanofibers was observed. 85% of the drug was released from the nanofibers after 4 h in simulated saliva solution suggesting that glutamine-loaded nanofibers have potential as an oromucosal drug delivery system [39].

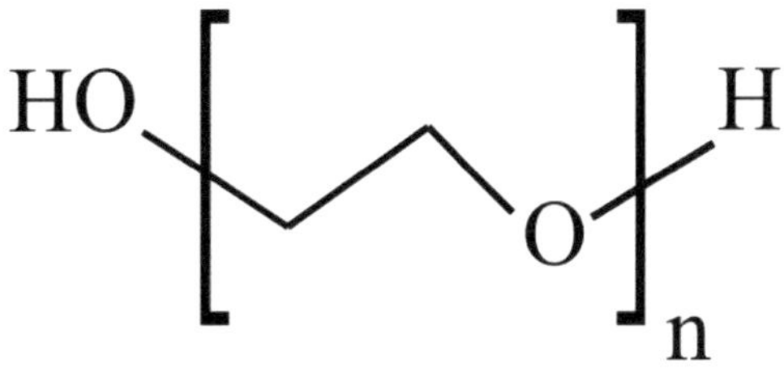

Figure 5.
Structural formula of poly(ethylene oxide).

Nanofiber-based local drug delivery system may be suitable for the treatment of cervical cancer. A pilot study by Zong et al. was carried out to examine the efficacy of cisplatin-loaded poly(ethylene oxide)/polylactide composite electrospun nanofibers as a local chemotherapy system against cervical cancer in mice via vaginal implantation. They have shown that a better balance between antitumor efficacy and systemic safety was achieved in a group of animals treated with nanofiber formulation as compared to i.v. injection group using an equal drug dose. Therefore, electrospun nanofibers present a promising approach to the local drug delivery via vaginal mucosa against cervical cancer [40].

4.1.4 Thiolated chitosan

Thiolated polymers are obtained by the addition of conjugated sulfhydryl groups. Thiolation of chitosan increases their mucoadhesive properties due to formation of disulfide bridges with cysteine domains of glycoproteins of the mucus. Moreover, chitosan and thiolated chitosan possess antiprotease activity due to their affinity to divalent cations, which are co-factors for proteases. All these characteristics make thiolated chitosan a promising material for mucosal administration of drugs, peptides, and proteins [13].

Leila Behbood et al. developed mucoadhesive nanofibers made up of thiolated chitosan as a drug delivery system for tetracycline and triamcinolone. Chitosan was modified via the immobilization of thiol groups from L-cysteine as a mucoadhesive reagent. Maximal mucoadhesion of nanofibers was observed at the pH value of 6. Release studies demonstrated that a sustained release of both drugs continued up to 48 h. The drug delivery system represented a novel tool for the improvement of therapeutic efficacy of various drugs that are poorly absorbed from different parts of the gastrointestinal tract. It was also shown to be an efficient system for treatment of oral ulceration [41].

The aim of the study performed by Samprasit et al. [42] was to fabricate mucoadhesive electrospun nanofiber mats containing α-mangostin for the maintenance of oral hygiene and reduction of the bacterial growth. Thiolated chitosan blended with polyvinyl alcohol was selected as the mucoadhesive polymer. The results of this study suggest that α-mangostin-loaded mucoadhesive electrospun nanofiber mats may be a promising material for the prevention of dental caries.

4.2 Transmucosal delivery of poorly soluble drugs using nanofibers

Many drugs are highly hydrophobic with poor water solubility and the number of poorly water-soluble drug candidates selected for development is rapidly increasing. It results in low oral absorption of these drugs as the absorption and bioavailability are limited by their poor solubility or slow dissolution in the gastrointestinal tract. This represents a major challenge for the pharmaceutical industry and novel formulation approaches are required. Several strategies including particle size reduction, micellization, salt formation, complexation, and solid dispersions have been developed to increase the oral absorption of such drugs. Solid dispersion is defined as the dispersion of one or more drugs in an inert matrix in the solid state. Simple eutectic mixtures, solid solutions, glass solutions of suspensions, and amorphous precipitates of a drug in a crystalline carrier are examples of solid dispersions [43]. Due to difficulties occurring during conventional methods of drug formulation, the applicability of solid dispersion systems has remained limited. Electrospun nanofibers provide a novel approach to improve the dissolution rate of even poorly water-soluble drugs, and thus might minimize the limitations of oral and oromucosal drug bioavailability [44].

As an example, maraviroc, an anti-HIV drug intended for intravaginal administration, was electrospun as solid dispersion made from either polyvinylpyrrolidone or poly(ethylene oxide) nanofibers or microfibers. In the study, the role of drug loading, distribution and crystallinity in determining drug release rates into aqueous media was investigated. It was shown that water-soluble electrospun materials can rapidly release maraviroc upon contact with moisture and that drug delivery is fast [45].

Salt formation improves drug solubility. However, drugs administered onto mucosal surfaces are effectively absorbed through mucosal surfaces if they are in the unionized form. Therefore, this strategy of enhanced drug dissolution is not advantageous for mucosal drug delivery.

The rate of dissolution of drugs formulated into particles is increased with their increasing surface area and decreasing particle size. Technologies used to decrease drug particle size to sub-micrometer range are being frequently applied to poorly water-soluble drug product development. Electrospinning is one of the technologies that can produce uniform nanosized polymeric nanofibers with drugs loaded into their structure. The release rates, and thus bioavailability of nanofiber-formulated drugs, are enhanced compared to those from the original drug substance [46].

Oral mucosa provides an interesting site of drug administration and absorption including poorly water-soluble drugs. However, they are not usually suitable for the formulation into classical oromucosal drug delivery systems. The formulation of nanofibers represents one possible route to achieve effective drug absorption via mucosal surfaces. Potrč et al. formulated polycaprolactone (PCL) nanofibers intended for oromucosal delivery of poorly water-soluble drugs. In this study, two model drugs, ibuprofen and carvedilol, with similar lipophilic properties, but differing in their molecular weights were chosen, and their influence on the nanofiber's physical properties and drug release profiles were investigated. The aim of the study was to establish a correlation between the drug's properties and the release characteristics of a PCL nanofiber-based delivery system. The results obtained in this study have shown that electrospinning can be used for the fabrication of drug-loaded PCL nanofibers with a high percentage of API embedded in them. The formulation of poorly water-soluble drugs into polycaprolactone-based nanofibers significantly increases their dissolution rate. However, the release rate of drugs from nanofibers is drug-dependent. Electrospinning was shown to be a very promising approach to the formulation of poorly water-soluble drugs in order to enhance their release and enable oromucosal administration [46].

4.3 Transmucosal delivery of macromolecules and nanoparticle-based vaccines using nanofibers

Mucosal surfaces are the most convenient routes for drug delivery to systemic circulation. However, transmucosal transport of macromolecular drugs such as peptides and proteins is much less effective as compared to low molecular weight drugs. Several strategies exploiting permeation enhancers, nanoparticulate carriers, nanofibers, and their combinations represent a promising strategy to facilitate transmucosal transport of macromolecules.

Recently, nanofiber-based mucoadhesive films have been invented for oromucosal administration of nanocarriers used for delivery of drugs and vaccines (**Figure 6**). The mucoadhesive film consists of an electrospun nanofibrous reservoir layer, a mucoadhesive film layer, and a protective backing layer. The mucoadhesive layer made of HPMC and Carbopol 934P polymers is responsible for tight adhesion of the whole system to the oral mucosa after application. The electrospun nanofibrous reservoir layer is intended to act as a reservoir for polymeric and lipid-based nanoparticles, liposomes, virosomes, virus-like particles, dendrimers and the like, plus macromolecular drugs, antigens and/or allergens. The extremely large surface area of nanofibrous reservoir layers allows high levels of nanoparticle loading. Nanoparticles can either be reversibly adsorbed to the surface of nanofibers or they can be deposited in the pores between the nanofibers. After mucosal application, nanofibrous reservoir layers are intended to promote prolonged release of nanoparticles into the submucosal tissue. Reversible adsorption of model nanoparticles as well as sufficient mucoadhesive properties was demonstrated. This novel system appears appropriate for the use in oral mucosa, especially for sublingual and buccal tissues [21].

Another example of novel multi-layered fibrous mucoadhesive film is based on self-assembled liposomes that are formed directly from nanofibrous layer after contacting with water. The idea came from a method of liposome preparation based on electrospinning technology. PVP was used as a nanofiber-forming matrix and phospholipid as liposome-forming molecules [47]. The membrane has been developed to improve the bioavailability of carvedilol. The whole system consists of an electrospun layer, an adhesive layer made of mucoadhesive film and a backing layer, similarly as previously described by Masek et al. [21]. Mucoadhesive film was formed using HPMC and CMC polymers and the standard solvent casting method. In general, this drug delivery system offered a novel platform for potential buccal delivery of drugs with a high first-pass effect.

One example for all macromolecular drugs is insulin. Insulin is a protein which is made of two polypeptide chains and it is not completely soluble in water. Many efforts have been made to find appropriate noninvasive routes of administration, including oral, pulmonary, rectal, oromucosal, and nasal. Although a certain degree of success exploiting all routes of mucosal administration of insulin was achieved, oromucosal, namely buccal and sublingual, delivery of insulin brings several

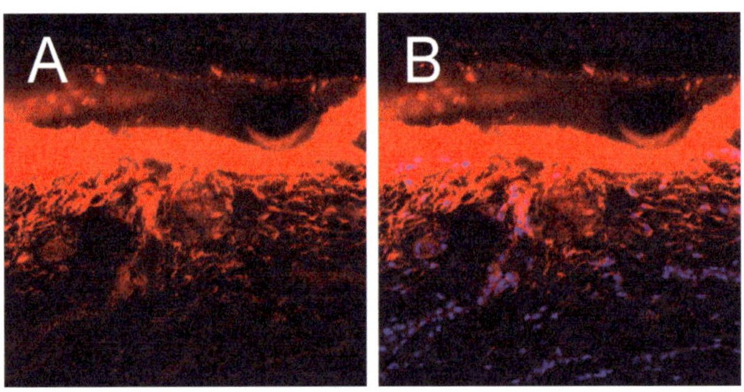

Figure 6.
Transmucosal penetration of model fluorescently labeled PLGA-PEG nanoparticles after ex vivo porcine mucosal administration using nanofibers. (A) PLGA-PEG nanoparticles (red) and (B) PLGA-PEG nanoparticles (red) plus nuclei (blue).

advantages. A number of attempts have been made to improve buccal insulin absorption by adding absorption enhancers or to modify the lipophilicity of insulin.

It is important to note that buccal and sublingual delivery of macromolecules including insulin using rodents as an animal model are of no value, because oral mucosa of rodents is highly keratinized. Therefore, the permeability for macromolecules and nanoparticles is negligible, while that of humans is quite high in the non-keratinized areas (sublingual and buccal). It means that the only animal models that can be of use when studying the human permeability of oral mucosa for macromolecules is pigs or dogs.

Several drug delivery systems have been tested for oromucosal insulin delivery, including sprays, mucoadhesive gels, and mucoadhesive films.

Transmucosal delivery of insulin via oral mucosa represents a novel approach. As an example, Sharma et al. have prepared an electrospun nanofiber-based membrane containing insulin molecules within the nanofiber structure. The solubility of insulin increases have been enhanced after formulation into polyvinyl alcohol (PVA) nanofibers as PVA itself is a surfactant and hence increases the solubility of insulin in the polymer solution. The release of insulin from a nanofiber membrane followed controlled release, and in vivo experiments confirmed high transmucosal delivery effectivity. Insulin release exhibits first order kinetics followed by an initial burst release necessary to produce the desired therapeutic activity. Furthermore, extremely high encapsulation efficacy of 99% of insulin indicates that nanofiber-based delivery system serves as an ideal carrier for the delivery of insulin via the sublingual route [48].

4.4 Fast-dissolving nanofiber-based drug delivery systems

Fast-dissolving drug delivery systems (FDDS) represent advanced formulations intended for oromucosal administration. FDDS are characterized by excellent flexibility and comfort for patients. The efficacy of drugs and rapid onset of their action are improved as FDDS dissolve within a minute in the oral cavity after the contact with saliva without the need of water for administration. FDDS are beneficial especially in pediatric and geriatric patients. It is also useful for delivery of drugs with local action.

Li et al. have fabricated nanofiber-based FDDS by electrospinning using PVA as the nanofiber-forming polymer and drug carrier. Caffeine and riboflavin were used as the model drugs. They found that drug release was completed in a burst manner. 100% of caffeine and 40% of riboflavin was dissolved within 60 s from the PVA nanofibrous matrices [49].

Isosorbide dinitrate-polyvinylpyrrolidone electrospun nanofibers were formulated and explored as a potentially sublingual membrane by Chen et al. The composition was favorable for the fabrication of the sublingual membrane as the dissolution was completed at 120 s. The pharmacokinetic study in rats demonstrated that the electrospinning fiber membrane had a higher C_{max} and lower T_{max} compared to the reference preparation [50].

Quan et al. demonstrated the concept of nanofiber-based FDDS also for poorly water-soluble drugs. In the study, feruloyl-oleyl-glycerol was used as a model drug and polyvinylpyrrolidone (PVP) K90 as a filament-forming polymer [51].

4.5 Mucosal administration sites of nanofiber-based drug delivery systems

Different mucosal sites of administration are a suitable target for nanofiber-based drug delivery systems, including oromucosal, nasal, vaginal, and ocular mucosa. Nanofiber-based drug delivery systems are used for both systemic and local drug administration.

4.5.1 Nanofibers in oromucosal drug delivery

Nanofiber membranes intended for oromucosal administration possess different properties according to the need of the desired indication and drug administered. The oromucosal site of administration, especially sublingual and buccal regions, are the most explored mucosal surfaces for drug delivery using the nanofiber-based system (**Figure 7**). The applications include fast-dissolving nanofiber-based formulations, mucoadhesive nanofibers, nanofiber-based formulations of poorly water-soluble drugs and, finally, nanofibers for delivery of different mucosal vaccines. Small drug molecules, macromolecules as therapeutic proteins, peptides, nucleic acids, and antigens are examples of the explored nanofiber-based systems intended for oromucosal administration. As description of these applications is broad, they are divided into relevant subchapters.

4.5.2 Nanofibers in vaginal drug delivery

Mucoadhesive nanofiber-based drug delivery systems are investigated for vaginal drug delivery. A wide range of materials have been explored for their fabrication into nanofibers. However, the local environment of vaginal surface has to be taken into the account when designing nanofiber-based vaginal drug delivery systems. Especially, the low pH values around pH 4.0 ± 0.5 make the difference as compared to other mucosal surfaces. As an example, progesterone-loaded drug delivery nanofiber constructs are described in Chapter 4.1.2.

4.5.3 Nanofibers in nasal drug delivery

Supramolecular peptide nanofibers have been explored as nasal formulation for vaccines and immunotherapy. Si et al. performed a study eliciting the immune response without the use of adjuvants and without measurable inflammation. Peptides comprise an epitope from influenza polymerase and the Q11 self-assembly domain formed nanofibers which were taken up by dendritic cells in lung-draining mediastinal lymph nodes after intranasal immunization. Nanofibers administered onto nasal mucosa elicited higher antigen-specific CD8+ T cell responses in the lung-draining lymph nodes as compared to subcutaneous immunizations, while retaining the noninflammatory character of the materials as opposed to other delivery sites. Influenza vaccines that can be administered intranasally or by other needle-free delivery routes have potential advantages over injected formulations

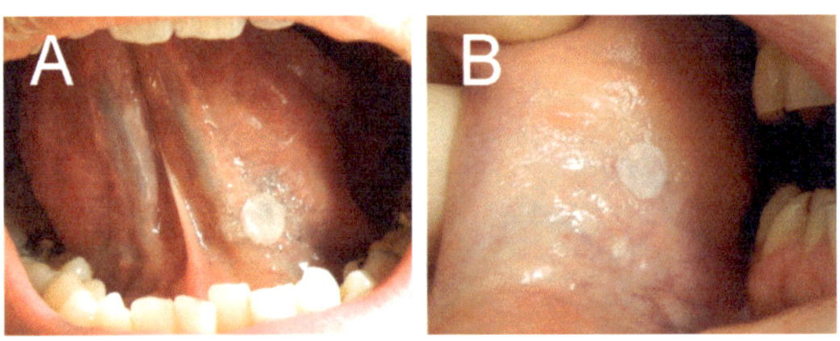

Figure 7.
Application of multi-layered mucoadhesive film with nanofibrous reservoir layer to sublingual (A) and buccal (B) mucosa.

in terms of patient compliance, cost, and ease of global distribution. It means that peptide nanofibers represent an interesting strategy for noninvasive influenza vaccines [52].

4.5.4 Nanofibers in ocular drug delivery

Ocular inserts are drug-impregnated formulations which can be placed onto ocular mucosa. Ocular inserts have been frequently used for reducing the frequency of administration, and, therefore, a controlled release profile is desired.

The objective of the study made by Mirzaeei et al. was to produce the electrospun nanofibers used as ophthalmic inserts. Triamcinolone acetonide was incorporated into a chitosan nanofiber-based ocular insert. This formulation increased the contact time between the drug and the conjunctival tissue, and thus decreased the number of administrations needed. This work showed that the concept of nanofibers in ophthalmic drug delivery is feasible [53].

5. Method of preparation and characterization of nanofibers for mucosal drug delivery

5.1 Fabrication of nanofibers for mucosal drug delivery

Nanofibers can be fabricated by several different techniques including drawing [54], phase separation [55], nanofiber seeding [56], template synthesis [57], self-assembly [58], etc. These techniques, on the other hand, allow neither control of nanofiber diameter nor continuous nanofiber production. Moreover, such techniques can only be used with specific polymers. On the contrary, electrospinning [59] is a resourceful and cost-effective technique that can be used to synthesize continuous nanofibers from numerous polymers and efficiently control their diameter.

Specifically, nanofibers produced by electrospinning (electrospun nanofibers) may be prepared from soluble polymers or from polymer solutions modified with additives such as particles, antimicrobial agents, or enzymes. Thanks to these additives, electrospun nanofibers may have desired properties. Therefore, electrospinning has gained a remarkable popularity in various disciplines boosting a recent steep rise in numbers of scientific publications.

Technically, electrospinning is a process that uses a strong electrical field to draw a polymer fluid into fine filaments. A typical electrospinning setup only requires a high voltage power supply, a syringe, a flat tip needle, and a conducting collector. When a polymer solution is charged with a high voltage, electrostatic force draws the fluid into a liquid jet (**Figure 8A**). Finally, solvent evaporation from the filaments results in solid nanofibers. In most cases, as-spun fibers deposit randomly on the electrode collector forming a nonwoven nanofiber mat. The basic equipment can be modified for various applications such as dual needle syringe (to make blended fibers), rotating collectors, etc.

Nanospider technology is a modern electrospinning technology for industrial-scale production of nanofibrous material without nozzles, needles, or spinnerets. Nanospider technology uses simply shaped electrodes covered by polymer solution (**Figure 8B**). It results in a mechanically simple technology with no parts that can be easily clogged (in comparison to needle-type electrospinning). Proven by an industrial operation, Nanospider technology provides high efficiency, outstanding fiber diameter, and web uniformity.

By electrospinning process it is possible to produce continuous nanofibers from a wide range of polymers. However, there are several parameters affecting the fiber

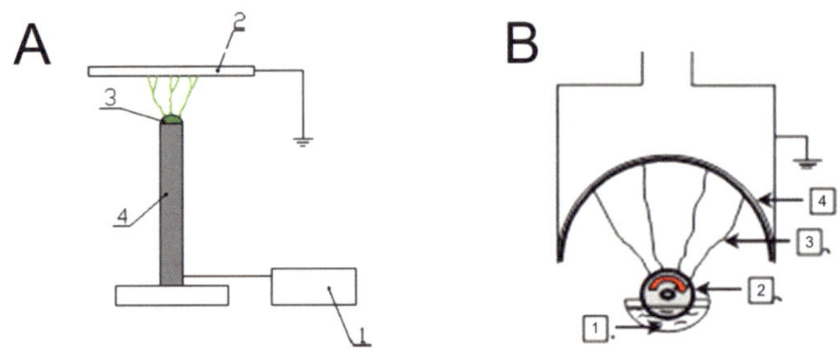

Figure 8.
Schematic representation of an electrospinning process (A): (1) a high voltage supply, (2) a grounded collector of nanofibers, (3) a polymer solution, and (4) a positive electrode. Schematic representation of Nanospider technology (B): (1) a polymer solution, (2) a rotating electrode with a high voltage supply, (3) created nanofibers, and (4) a grounded collector of nanofibers.

morphology and properties of electrospun nanofibers. The whole process can be controlled by four important characteristics: (i) process parameters such as voltage, spinning distance, flow rate, or collecting plate, (ii) systemic and (iii) solution parameters which affected concentration, conductivity, or surface tension of a polymer solution, and (iv) physical parameters such as humidity, temperature, or air velocity. All mentioned parameters are major factors affecting the fiber morphology and web properties. Because these variables interrelate, a small change in either of these variables can have a significant impact on nanofiber morphology or even the electrospinning process altogether. Solvents or their mixtures used for dissolving of polymer have a direct impact on the electrospinning process and morphology of the resulting nanofibers. Laboratory experience has shown that a solvent that creates mostly 80–99 wt% of polymer solution has a dominant impact. Solvents primarily determine (i) conformation of dissolved macromolecular chains, (ii) easiness of charging the surface layer, (iii) cohesion of the polymer solution due to the surface tension forces, and (iv) the rate of solidification of the liquid jet during evaporation of the solvent.

5.2 Materials suitable for nanofiber-based mucosal drug delivery systems

Nanofiber-based mucosal drug delivery systems cannot be in general electrospun from any polymer as they require specific properties. As examples of very interesting materials for nanofiber-based mucosal drug delivery systems, the following can be mentioned: biopolymers such as gelatin [60], chitosan [31], collagen [61], cellulose [62], silk fibroin [63], hyaluronic acid [64], polylactic acid [65], or polycaprolactone [46].

In addition to specific materials used for production of nanofiber mucosal drug delivery systems they also require surface modification of nanofibers. After the functionalization of a nanofiber surface, drugs might be bound or conjugated to nanofiber surfaces. In such a way, the release of drugs would be attenuated, and the functionality of the surface-immobilized biomolecules could be preserved. This strategy is usually applied in order to overcome the issue of initial burst release as well as short release time. The most used surface modifications are: (i) plasma treatment, (ii) wet chemical method, or (iii) co-electrospinning of active agents.

Different sources of plasma (e.g., oxygen, argon, ammonia, air) used for treatment of nanofibers can create different functional groups (such as carboxyl or amine groups) on the nanofiber surface. This kind of chemical groups may interact with particular drugs and create covalent bonds. However, if a target biomolecule is chemically bound onto the nanofiber surface, it would hardly be released.

Therefore, this technique is more suitable for drugs, where a slow and prolonged release of the agent is required. Plasma treatment can also change hydrophilicity and hydrophobicity of nanofibers.

Wet chemical method allows changing the wettability of nanofibers under acidic or basic conditions. Surface of nanofibers deep in mesh can also be modified by the wet chemical method. Plasma treatment is, on the contrary, more suitable for flat materials.

By co-electrospinning of active agents, it is possible to directly expose biological functional agents on the surface of nanofibers. Conjugating the biomolecules (DNA, growth factors, or enzymes) to the fiber surfaces allows their slow release into a nearby tissue significantly preserving the functionality of biomolecules.

Functionalization of the nanofiber surface enables loading of drugs. There are many methods how to load them. The most popular and used techniques are: (i) physical adsorption, (ii) nanoparticle assembly method, (iii) layer by layer method, and (iv) chemical immobilization.

In the case of physical adsorption, there is no need for nanofiber functionalization after electrospinning. The fiber web is simply immersed into a solution containing drugs and dried afterward. The same method can be used in the case of nanoparticles containing biological agents. Chemical immobilization requires functionalized surface of nanofibers by the plasma treatment or chemical wet method. Afterward, functional groups on the surface of nanofibers chemically react with added drugs and create covalent bonds. By the multilayer method, it is possible to produce a nanofiber sandwich with different properties on both surfaces. After electrospinning of one layer with drugs added during the electrospinning process, a sandwich with another nanofiber layer without drugs can be created.

5.3 Methods for characterization of nanofibers for mucosal drug delivery

Biomedical applications of nanofibers such as the mucosal drug delivery system put special requirements on the three-dimensional electrospun materials. Besides the biocompatibility, the morphology of nanofibers is one of the most important attributes. The specific surface area, volume, and the size of the pores have considerable effect on the loading capacity of drugs. The following methods are used to characterize electrospun materials for mucosal drug delivery systems.

Imaging methods are used for evaluation of nanofiber structure. Imaging methods involve scanning electron microscopy (SEM), transmission electron microscopy (TEM), and atomic force microscopy (AFM). By SEM and TEM, it is possible to evaluate the nanofiber orientation, nanofiber diameter, and the morphology of nanofibers, which do not only affect the mechanical properties of electrospun materials, but also play a key role in the loading capacity of drugs in the mucosal drug delivery systems. Imaging methods also allow visualization of the morphology of nanofibers at various points of an electrospun material.

Loading of drugs must be controlled during the assessment of biological properties and this ability is significantly affected by the physical properties such as pore size and volume of electrospun material. The surface area and the porosity could be measured by mercury porosimetry or by Brunauer-Emmett-Teller (BET) surface area analysis. A pore size distribution is one of the most often presented results of mercury porosimetry. However, the mercury porosimetry can produce a misleading result due to the mechanical deformation of the nanofibers [66]. To overcome this issue, BET measurements are used to measure the specific surface area value and distribution of pores.

Besides the morphology of nanofibers, the chemical composition is an important attribute for materials applicable in the mucosal drug delivery systems. The

Fourier-transform infrared spectroscopy (FTIR), differential scanning calorimetry (DSC) [67], and differential thermal analysis (DTA) are essential methods for measuring the chemical composition. These methods allow detection of abundance of each polymer in the final product. FTIR indicates degradation of nanofibers (for biodegradable materials) as well as it may show their bioactivity. The bioactivity is detected by infrared spectra obtained via FTIR that identifies the functional chemical groups. The hydrophilic/hydrophobic character of electrospun materials influences the loading capacity of nanofibers as well. To determine the degree of hydrophilicity, contact angle measurement is one of the most used methods.

The last most important and crucial characteristic of nanofibers is the release of drugs from electrospun materials [68]. As it was mentioned above, slow or fast release of target drugs might be changed by a different surface functionalization of nanofibers. For this purpose, a dissolution testing apparatus with UV-Vis spectrophotometer is essential to control the release profile.

6. Conclusions and future directions

Different parts of the human body are covered by mucosa with different features, barrier properties for drug delivery, and also with different accessibility. Formulation of drugs into nanofibers represents one of the new trends in mucosal drug delivery. Due to their extremely large surface area, unique surface topology and porosity, nanofiber-based drug delivery systems enable transmucosal delivery of poorly water-soluble drugs, macromolecules, nanoparticles, and vaccine delivery carriers.

Extraordinary flexibility of nanofibers enables us to follow unique anatomical specialities of mucosal surfaces, and hence helps to overcome different absorption barriers of mucosal sites. Moreover, the flexibility of nanofibers helps to significantly increase the comfort of nanofiber-based drug delivery formulations for patients.

Mucoadhesive nanofibers with drug-controlled release properties and nanofibers with extremely fast-dissolving properties are examples of a great variety of nanofiber-based materials and also examples of a variety of drug delivery system properties advantageous for mucosal administration. Different mucosal sites of administration, including sublingual, buccal, nasal, vaginal, and ocular mucosa, are suitable targets for nanofiber-based drug delivery systems. Mucosal surfaces, as a portal of entry of various infectious pathogens, naturally possess great potential for induction of defensive immune responses against such pathogens. Nanofiber-based delivery platforms, owning their unique properties, may play an important role in formulation of antigens into next-generation vaccine delivery systems intended for mucosal administration.

Technologies of electrospinning, such as Nanospider technology, are modern electrospinning technologies enabling cost-effective industrial-scale production of nanofibrous materials, among others, suitable for mucosal drug delivery applications.

Combinations of nanofiber-based formulations and chemical enhancers have a great potential to increase penetration of drug molecules and nanoparticle-based drug and vaccine delivery systems. In conclusion, nanofibers represent a new emerging trend in formulation of drug and vaccine delivery systems for mucosal administration.

Acknowledgements

This work was supported by the Ministry of Education, Youth, and Sports OPVVV project "CEREBIT" CZ.02.1.01/0.0/0.0/16_025/0007397 (MR,JT); OPVVV project "FIT" CZ.02.1.01/0.0/0.0/15_003/0000495 (JT); the Ministry of Health CZ AZV-ČR 15-32198A (MR, JT), the project MZE0002716202, RO0518 of the Czech Ministry of Agriculture (JT).

Conflict of interest

There are no conflicts to declare.

Author details

Josef Mašek[1*], Eliška Mašková[1,2], Daniela Lubasová[3], Roman Špánek[3], Milan Raška[1,4] and Jaroslav Turánek[1]

1 Department of Pharmacology and Immunotherapy, Veterinary Research Institute, Brno, Czech Republic

2 Department of Pharmaceutics, Faculty of Pharmacy, University of Veterinary and Pharmaceutical Sciences Brno, Brno, Czech Republic

3 Institute for Nanomaterials, Advanced Technologies and Innovation, Technical University of Liberec, Liberec, Czech Republic

4 Department of Immunology and Institute of Molecular and Translational Medicine, Faculty of Medicine and Dentistry, Palacky University Olomouc, Olomouc, Czech Republic

*Address all correspondence to: masek@vri.cz

References

[1] Harris D, Robinson JR. Drug delivery via the mucous membranes of the oral cavity. Journal of Pharmaceutical Sciences. 1992;**81**(1):1-10. DOI: 10.1002/jps.2600810102

[2] Squier CA, Wertz P. Structure and function of the oral mucosa and implications for drug delivery. In: Rathbone MJ, editor. Oral Mucosal Drug Delivery. New York: Taylor & Francis; 1996. pp. 1-26

[3] Salamat-Miller N, Chittchang M, Johnston TP. The use of mucoadhesive polymers in buccal drug delivery. Advanced Drug Delivery Reviews. 2005;**57**:1666-1691. DOI: 10.1016/j.addr.2005.07.003

[4] Patel VF, Liu F, Brown MB. Advances in oral transmucosal drug delivery. Journal of Controlled Release. 2011;**153**:106-116. DOI: 10.1016/j.jconrel.2011.01.027

[5] Hillery AM, Park K. Drug Delivery: Fundamentals and Applications. 2nd ed. Boca Raton, USA: CRC Press by Taylor & Francis Group, LLC; 2017. p. 632

[6] Allen A, Bell A, McQueen S. Mucus and mucosal protection. In: Allen A, Flemstrom M, Garner A, Silen W, Turnberg LA, editors. Mechanisms of Mucosal Protection in the Upper Gastrointestinal Tract. New York: Raven Press; 1984. pp. 195-202

[7] Peppas NA, Buri PA. Surface, interfacial, and molecular aspects of polymer bioadhesion on soft tissues. Journal of Controlled Release. 1985;**2**:257-275. DOI: 10.1016/0168-3659(85)90050-1

[8] Sarkar MA. Drug metabolism in the nasal mucosa. Pharmaceutical Research. 1992;**9**:1-9. DOI: 10.1023/A:1018911206646

[9] Ugwoke MI, Agu RU, Verbeke N, Kinget R. Nasal mucoadhesive drug delivery: Background, applications, trends and future perspectives. Advanced Drug Delivery Reviews. 2005;**57**:1640-1665. DOI: 10.1016/j.addr.2005.07.009

[10] Kaliner M, Marom Z, Patow C, Shelhamer J. Human respiratory mucus. The Journal of Allergy and Clinical Immunology. 1984;**73**:318-323

[11] Caramella CM, Rossi S, Ferrari F, Bonferoni MC, Sandri G. Mucoadhesive and thermogelling systems for vaginal drug delivery. Advanced Drug Delivery Reviews. 2015;**92**:39-52. DOI: 10.1016/j.addr.2015.02.001

[12] Chappell CA, Rohan LC, Moncla BJ, Wang L, Meyn LA, Bunge K, et al. The effect of reproductive hormones on the physical properties of cervicovaginal fluid. American Journal of Obstetrics and Gynecology. 2014;**211**:226.e1-226.e7. DOI: 10.1016/j.ajog.2014.03.041

[13] Rathbone MJ, Şenel S, Pather I. Oral Mucosal Drug Delivery and Therapy. Advances in Delivery Science and Technology. New York: Springer; 2015. DOI: 10.1007/978-1-4899-7558-4

[14] Gandhi RB, Robinson JR. Oral cavity as a site for bioadhesive drug delivery. Advanced Drug Delivery Reviews. 1994;**13**:43-74. DOI: 10.1016/0169-409X(94)90026-4

[15] Siegel IA, Hall SH, Stambaugh R. Permeability of the oral mucosa. In: Squier CA, Meyer J, editors. Current Concepts of the Histology of Oral Mucosa. Springfield, IL: Carles Thomas; 1971. pp. 274-286

[16] Illum L. Nasal drug delivery—Possibilities, problems and solutions. Journal of Controlled Release.

2003;**87**:187-198. DOI: 10.1016/S0168-3659(02)00363-2

[17] Brandtzaeg P. Induction of secretory immunity and memory at mucosal surfaces. Vaccine. 2007;**25**:5467-5484. DOI: 10.1016/j.vaccine.2006.12.001

[18] Elson CO, Ealding W. Cholera toxin feeding did not induce oral tolerance in mice and abrogated oral tolerance to an unrelated protein antigen. Journal of Immunology. 1984;**133**:2892-2897

[19] Mutsch M, Zhou W, Rhodes P, Bopp M, Chen RT, Linder T, et al. Use of the inactivated intranasal influenza vaccine and the risk of Bell's palsy in Switzerland. New England Journal of Medicine. 2004;**350**:896-903. DOI: 10.1056/NEJMoa030595

[20] Czerkinsky C, Cuburu N, Kweon MN, Anjuere F, Holmgren J. Sublingual vaccination. Human Vaccines. 2011;**7**:110-114

[21] Masek J, Lubasova D, Lukac R, Turanek-Knotigova P, Kulich P, Plockova J, et al. Multi-layered nanofibrousmucoadhesive films for buccal and sublingual administration of drug-delivery and vaccination nanoparticles—Important step towards effective mucosal vaccines. Journal of Controlled Release. 2017;**249**:183-195. DOI: 10.1016/j.jconrel.2016.07.036

[22] Merkus FWHM, Schipper NG, Hermens WAJJ, Romeijn SG, Verhoef JC. Absorption enhancers in nasal drug delivery: Efficacy and safety. Journal of Controlled Release. 1993;**24**(1-3):201-208. DOI: 10.1016/0168-3659(93)90179-9

[23] Şenel S, Hincal AA. Drug permeation enhancement via buccal route: Possibilities and limitations. Journal of Controlled Release. 2001;**72**(1-3):133-144. DOI: 10.1016/S0168-3659(01)00269-3

[24] Ganem-Quintanar A, Kalia YN, Falson-Rieg F, Buri P. Mechanisms of oral permeation enhancement. International Journal of Pharmaceutics. 1997;**156**:127-142. DOI: 0.1016/S0378-5173(97)00193-2

[25] Zhang H, Zhang J, Streisand JB. Oral mucosal drug delivery: Clinical pharmacokinetics and therapeutic applications. Clinical Pharmacokinetics. 2002;**41**(9):661-680. DOI: 10.2165/00003088-200241090-00003

[26] Bernkop-Schnürch A, Hornof M. Intravaginal drug delivery: Design, challenges and solutions. American Journal of Drug Delivery. 2003;**1**:241-254. DOI: 10.2165/00137696-200301040-00003

[27] Lowry D. Delivery of subunit vaccines. In: Foged C, Rades T, Perrie Y, Hook S, editors. Subunit Vaccine Delivery. New York, NY: Springer; 2014. pp. 331-346. DOI: 10.1007/978-1-4939-1417-3

[28] Song JH, Nguyen HH, Cuburu N, Horimoto T, Ko SY, Park SH, et al. Sublingual vaccination with influenza virus protects mice against lethal viral infection. Proceedings of the National Academy of Sciences of the United States of America. 2008;**105**:1644-1649. DOI: 10.1073/pnas.0708684105

[29] Cuburu N, Kweon MN, Song JH, Hervouet C, Luci C, Sun JB, et al. Sublingual immunization induces broad-based systemic and mucosal immune responses in mice. Vaccine. 2007;**25**:8598-8610

[30] Yuki Y, Kiyono H. Mucosal vaccines: Novel advances in technology and delivery. Expert Review of Vaccines. 2009;**8**:1083-1097. DOI: 10.1586/erv.09.61

[31] Lancina MG, Shankar RK, Yang H. Chitosan nanofibers for transbuccal insulin delivery. Journal of

Biomedical Materials Research, Part A. 2017;**105**(5):1252-1259. DOI: 10.1002/jbm.a.35984

[32] Wongsasulak S, Pathumban S, Yoovidhya T. Effect of entrapped-tocopherol on mucoadhesivity and evaluation of the release, degradation, and swelling characteristics of zein-chitosan composite electrospun fibers. Journal of Food Engineering. 2014;**120**:110-117. DOI: 10.1016/j.jfoodeng.2013.07.028

[33] Mendes AC, Sevilla Moreno J, Hanif M, TEL Douglas, Chen M, Chronakis IS: Morphological, mechanical and mucoadhesive properties of electrospun chitosan/phospholipid hybrid nanofibers. International Journal of Molecular Sciences 2018;19(8). pii: E2266. DOI: 10.3390/ijms19082266

[34] Dilamian M, Montazer M, Masoumi J. Antimicrobial electrospun membranes of chitosan/poly(ethylene oxide) incorporating poly(hexamethylene biguanide) hydrochloride. Carbohydrate Polymers. 2013;**94**(1):364-371. DOI: 10.1016/j.carbpol.2013.01.059

[35] Brako F, Thorogate R, Mahalingam S, Raimi-Abraham B, Craig DQM, Edirisinghe M. Mucoadhesion of progesterone-loaded drug delivery nanofiber constructs. ACS Applied Materials & Interfaces. 2018;**10**(16):13381-13389. DOI: 10.1021/acsami.8b03329

[36] Brako F, Raimi-Abraham BT, Mahalingam S, Craig DQ, Edirisinghe M. The development of progesterone-loaded nanofibers using pressurized gyration: A novel approach to vaginal delivery for the prevention of pre-term birth. International Journal of Pharmaceutics. 2018;**540**:31-39. DOI: 10.1016/j.ijpharm.2018.01.043

[37] Nazari K, Kontogiannidou E, Ahmad RH, Andreadis D, Rasekh M, Bouropoulos N, et al. Fibrous polymeric buccal film formulation, engineering and bio-interface assessment. European Polymer Journal. 2017;**97**:147-157. DOI: 10.1016/j.eurpolymj.2017.09.046

[38] El-Newehy Mohamed H, El-Naggar Mehrez E, Saleh A, Hany E-H, Meera M, Salem A-D. Green electrospining of hydroxypropyl cellulose nanofibres for drug delivery applications. Journal of Nanoscience and Nanotechnology. 2018;**18**(2):805-814. DOI: 10.1166/jnn.2018.13852

[39] Tort S, Acartürk F. Preparation and characterization of electrospun nanofibers containing glutamine. Carbohydrate Polymers. 2016;**152**:802-814. DOI: 10.1016/j.carbpol.2016.07.028

[40] Zong S, Wang X, Yang Y, Wu W, Li H, Ma Y, et al. The use of cisplatin-loaded mucoadhesive nanofibers for local chemotherapy of cervical cancers in mice. European Journal of Pharmaceutics and Biopharmaceutics. 2015;**93**:127-135. DOI: 10.1016/j.ejpb.2015.03.029

[41] Behbood L, Karimi S, Mirzaei E, Mohammadi G, Azami M, Arkan E. Mucoadhesive chitosan electrospun nanofibers containing tetracycline and triamcinolone as a drug delivery system. Fibers and Polymers. 2018;**19**:1454-1462. DOI: 10.1007/s12221-018-8087-1

[42] Samprasit W, Rojanarata T, Akkaramongkolporn P, Ngawhirunpat T, Kaomongkolgit R, Opanasopit P. Fabrication and *in vitro*/*in vivo* performance of mucoadhesive electrospun nanofiber mats containing α-mangostin. AAPS PharmSciTech. 2015;**16**(5):1140-1152. DOI: 10.1208/s12249-015-0300-6

[43] Ignatious F, Sun L, Lee CP, Baldoni J. Electrospun nanofibers in oral drug delivery. Pharmaceutical Research. 2010;**27**(4):576-588. DOI: 10.1007/s11095-010-0061-6

[44] Yu DG, Zhu LM, White K, Branford-White C. Electrospun nanofiber-based drug delivery systems. Health. 2009;**1**:67-75. DOI: 10.4236/health.2009.12012

[45] Ball C, Woodrow KA. Electrospun solid dispersions of maraviroc for rapid intravaginal preexposure prophylaxis of HIV. Antimicrobial Agents and Chemotherapy. 2014;**58**(8):4855-4865. DOI: 10.1128/AAC.02564-14

[46] Potrč T, Baumgartner S, Roškar R, Planinšek O, Lavrič Z, Kristl J, et al. Electrospun polycaprolactone nanofibers as a potential oromucosal delivery system for poorly water-soluble drugs. European Journal of Pharmaceutical Sciences. 2015;**75**:101-113. DOI: 10.1016/j.ejps.2015.04.004

[47] Chen J, Pan H, Yang Y, Xiong S, Duan H, Yang X, et al. Self-assembled liposome from multi-layered fibrous mucoadhesive membrane for buccal delivery of drugs having high first-pass metabolism. International Journal of Pharmaceutics. 2018;**547**(1-2):303-314. DOI: 10.1016/j.ijpharm.2018.05.062

[48] Sharma A, Gupta A, Rath G, Goyal A, Mathur RB, Dhakate SR. Electrospun composite nanofiber-based transmucosal patch for anti-diabetic drug delivery. Journal of Materials Chemistry B. 2013;**1**(27):3410-3418. DOI: 10.1039/C3TB20487A

[49] Li X, Kanjwal MA, Lin L, Chronakis IS. Electrospun polyvinyl-alcohol nanofibers as oral fast-dissolving delivery system of caffeine and riboflavin. Colloids and Surfaces B: Biointerfaces. 2013;**103**:182-188. DOI: 10.1016/j.colsurfb.2012.10.016

[50] Chen J, Wang X, Zhang W, Yu S, Fan J, Cheng B, et al. A novel application of electrospinning technique in sublingual membrane: Characterization, permeation and *in vivo* study. Drug Development and Industrial Pharmacy. 2016;**42**:1365-1374. DOI: 10.3109/03639045.2015.1135939

[51] Quan J, Yu Y, Branford-White C, Williams GR, Yu DG, Nie W, et al. Preparation of ultrafine fast-dissolving feruloyl-oleyl-glycerol-loaded polyvinylpyrrolidone fiber mats via electrospinning. Colloids and Surfaces B: Biointerfaces. 2011;**88**(1):304-309. DOI: 10.1016/j.colsurfb.2011.07.006

[52] Si Y, Wen Y, Kelly SH, Chong AS, Collier JH. Intranasal delivery of adjuvant-free peptide nanofibers elicits resident CD8 ı T cell responses. Journal of Controlled Release. 2018;**282**:120-130. DOI: 10.1016/j.jconrel.2018.04.031

[53] Mirzaeei S, Berenjian K, Khazaei R. Preparation of the potential ocular inserts by electrospinning method to achieve the prolong release profile of triamcinolone acetonide. Advanced Pharmaceutical Bulletin. 2018;**8**(1):21-27. DOI: 10.15171/apb.2018.003

[54] Xing X, Wang Y, Li B. Nanofibers drawing and nanodevices assembly in poly(trimethylene terephthalate). Optics Express. 2008;**16**(14):10815-10822. DOI: 10.1364/OE.16.016815

[55] Katsogiannis KAG, Vladisavljević GT, Georgiadou S. Porous electrospun polycaprolactone (PCL) fibres by phase separation. European Polymer Journal. 2015;**69**:284-295. DOI: 10.1016/j.eurpolymj.2015.01.028

[56] Zhang X, Goux JW, Manohar SK. Synthesis of polyaniline nanofibers by "nanofiber seeding". Journal of the American Chemical Society. 2004;**126**(14):4502-4503. DOI: 10.1021/ja031867a

[57] Wang Y, Zheng M, Lu H, Feng S, Ji G, Cao J. Template synthesis of carbon nanofibers containing linear mesocage arrays. Nanoscale Research Letters. 2010;**5**(6):913-916. DOI: 10.1007/s11671-010-9562-9

[58] Rolandi M, Rolandi R. Self-assembled chitin nanofibers and applications. Advances in Colloid and Interface Science. 2014;**207**:216-222. DOI: 10.1016/j.cis.2014.01.019

[59] Fong H, Reneker DH. Electrospinning and formation of nanofibers. In: Structure Formation in Polymeric Fibers. Munich: Hanser; 2001. pp. 225-246

[60] Li H, Wang M, Williams GR, Wu J, Sun X, Lv Y, et al. Electrospun gelatin nanofibers loaded with vitamins A and E as antibacterial wound dressing materials. RSC Advances. 2016;**6**: 50267-50277. DOI: 10.1039/C6RA05092A

[61] Hall Barrientos IJ, Paladino E, Szabó P, Brozio S, Hall PJ, Oseghale CI, et al. Electrospun collagen-based nanofibres: A sustainable material for improved antibiotic utilisation in tissue engineering applications. International Journal of Pharmaceutics. 2017;**531**(1):67-79. DOI: 10.1016/j.ijpharm.2017.08.071

[62] Nazari K, Kontogiannidou E, Ahmad RH, Gratsani A, Rasekh M, Arshad MS, et al. Development and characterisation of cellulose based electrospun mats for buccal delivery of non-steroidal anti-inflammatory drug (NSAID). European Journal of Pharmaceutical Sciences. 2017;**102**:147-155. DOI: 10.1016/j.ejps.2017.02.033

[63] Sheikh FA, Ju HW, Lee JM, Moon BM, Park HJ, Lee OJ, et al. 3D electrospun silk fibroin nanofibers for fabrication of artificial skin. Nanomedicine: Nanotechnology, Biology and Medicine. 2015;**11**(3):681-691. DOI: 10.1016/j.nano.2014.11.007

[64] Uppal R, Ramaswamy GN, Arnold C, Goodband R, Wang Y. Hyaluronic acid nanofiber wound dressing—Production, characterization, and *in vivo* behavior. Journal of Biomedical Materials Research. Part B, Applied Biomaterials. 2011;**97**(1):20-29. DOI: 10.1002/jbm.b.31776

[65] Gómez-Pachón EY, VeraGraziano R, Campos RM. Structure of poly (lactic-acid) PLA nanofibers scaffolds prepared by electrospinning. IOP Conference Series: Materials Science and Engineering. 2014;**59**:1-9. DOI: 10.1088/1757-899X/59/1/012003

[66] Pham QP, Sharma U, Mikos AG. Electrospun poly(ε-caprolactone) microfiber and multilayer nanofiber/microfiber scaffolds: Characterization of scaffolds and measurement of cellular infiltration. Biomacromolecules. 2006;7:2796-2805. DOI: 10.1021/bm060680j

[67] Paaver U, Heinämäki J, Laidmäe I, Lust A, Kozlova J, Sillaste E, et al. Electrospun nanofibers as a potential controlled-release solid dispersion system for poorly water-soluble drugs. International Journal of Pharmaceutics. 2015;**479**(1):252-260. DOI: 10.1016/j.ijpharm.2014.12.024

[68] Weng L, Xie J. Smart electrospun nanofibers for controlled drug release: Recent advances and new perspectives. Current Pharmaceutical Design. 2015;**21**(15):1944-1959. DOI: 10.2174/1381612821666150302151959

Chapter 11

Stupendous Nanomaterials: Carbon Nanotubes Synthesis, Characterization, and Applications

Kalaiselvan Shanmugam, J. Manivannan and M. Manjuladevi

Abstract

Carbon nanotubes are promising to revolutionize several fields in material science and are suggested to open the way into nanotechnology. These circular rod-shaped carbon nanostructures have novel characteristics that lead them to being potentially beneficial in many applications in nanoscience and nanotechnology. Their precise surface place, stiffness, power, and resilience have brought about lots of exhilaration in various areas. Nanotubes are categorized as single-walled nanotubes, double-walled nanotubes, and multi-walled nanotube. Various techniques have been evolved to produce nanotubes in bulk, including of arc discharge, laser ablation, chemical vapor deposition, electrolysis, and ball milling. Since their first observation nearly 20 years ago by Iijima, carbon nanotubes have been the focus of considerable research. Numerous researchers have reported remarkable physical and chemical properties for this new form of advanced carbon nanomaterials. Carbon nanotubes offer tremendous opportunities for the development of new material systems. This paper provides a concise report on recent advances in carbon nanotubes and their potential applications.

Keywords: carbon nanotubes, SWCNTs, MWCNTs, CVD, spray pyrolysis, SEM, HRTEM, purification

1. Introduction

Up until 1985, it was generally accepted that solid elemental carbon occurs in two different crystalline states: diamond and graphite. In the structure of diamond, every atom is tetrahedrally encircled by four sp^3 covalently neighboring carbon atoms. It brings about a special system of carbon developed on a cube-like face focused lattice. The structure of graphite comprises of graphene layers inside which the sp^2 neighboring carbon atoms form a planar polygon honeycomb arrangement. The bonding of carbon atoms in a graphene plane is through very strong valence bonds though the holding between two graphene layers is feeble Vander Waals bonds. In 1985, vital revolution in carbon research was realized by the work of Kroto et al [1] that resulted in the discovery of an enormous family of all carbon molecules, referred to as fullerenes. The fullerenes are closed cage carbon molecules formed by 12 pentagonal rings and 20 hexagonal rings. The molecular structure of

 IntechOpen

C60 shows that each of its pentagon is enclosed by 5 hexagons. The crucial particularity is the presence of five-membered rings, which help the arc form a confined cage structure. In 1990, Kratschmer et al. [2] found that the residue made by arcing carbon electrodes contained C60 and different fullerenes. A year after multi-walled carbon nanotubes (MWNT), hollow long thin cylinders of carbon consisting of concentric graphite layers with diameters in the nanometer range and consisting of carbon atoms were primarily observed in 1991 by Sumio Iijima at the NEC research lab, when he studied the carbon residue made up of by-products obtained throughout the synthesis of fullerenes by the electrical arc discharge technique.[3] from that point after 2 years, single-walled carbon nanotubes (SWNTs) were produced. Ijima along with Ichihasi [4] used carbon electrodes with a small amount of iron and filled the chamber around the carbon arc with methane and argon gas, which yielded the single-walled carbon nanotubes. [5]. In 1996, Smalley synthesized bundles of single-walled carbon nanotubes for the first time [6]. These miracle materials have remarkable properties such as excellent tensile strength, thermal conductivity, and electrical conductivity that make them potentially useful in many applications in nanotechnology, electronics, optics, and composites.

2. Allotropes of carbon

Carbon, the sixth element in the periodic table, is the most versatile element in terms of the variety of materials it may form. Each carbon atom has four electrons in its valence shell ($2s^2 2p^2$) that can be used to form covalent bonds with other atoms. With different hybridizations, carbon atoms can be arranged into materials with different structures, such as diamond, graphite, fullerene, and CNTs. Diamond and graphite are two of the most common forms in the carbon family. In diamond, the sp^3-hybridized carbon atoms are covalently bonded to four neighboring carbon atoms and therefore have tetrahedral structure. Due to the strength of the carbon–carbon bond and its three-dimensional structure, diamond is the hardest known material. Since there are no free electrons in diamond, it is an electric insulator. Graphite, on the other hand, is composed of sp^2-hybridized carbon atoms. Each of these sp^2-hybridized carbon molecules is attached to three other carbon atoms, departing one free electron in a p-π orbital. This gives graphite a two-dimensional layer-like structure. The π electrons are delocalized inside the layers, which furnish the graphite with superb thermal and electrical conductivity. Notwithstanding, this sort of layered structure makes graphite exceptionally delicate, because the layers can slide in analogous direction with ease. Other important members in the carbon family are the fullerenes that were discovered in 1985. Fullerenes are closed convex cage molecules made of sp^2-hybridized carbon atoms. Each of the carbon atoms is connected to three neighboring carbon atoms to form only pentagonal and hexagonal faces. C60, which looks like a soccer ball, is the simplest fullerene molecule containing 12 pentagons and 10 hexagons[7] (**Table 1**).

2.1 Structure of carbon nanotubes

CNTs can be regarded as giant fullerenes, with a one-dimensional tubular structure. CNTs have a high aspect ratio and their length can be millions of times greater than their tiny diameter. All carbon atoms in CNTs are sp^2 hybridized and connected with each other to form six-member rings that uniformly cover the surface of the CNT sidewall. Unlike graphite, there are no dangling carbon bonds in a perfect CNT. CNTs are classified as single-walled carbon nanotubes (SWNTs) and multi-walled carbon nanotubes (MWNTs) according to the number of graphene layers.

DOI: http://dx.doi.org/10.5772/intechopen.90318

Property	Graphite	Diamond	Fullerenes (bucky ball)	Carbon nanotubes
Color	Steel black to gray	Colorless	Black solid/magenta in solution	Black
Crystal structure	Tabular	Cubic	Truncated icosahedron	Cylindrical
Density (g/cm^3)	1.9–2.3	3.515	1.69	1.33–1.4
Melting point (°C)	3652–3297	3550	>800	Similar to graphite
Boiling point (°C)	4200	4827	NA	NA
Hybridization	Sp^2—trigional planar	Sp^3—tetrahedral	Sp^2—trigional planar	Sp^2—trigional planar
Bond angle	120°	109°5′	120°	NA
Nature of bonds in structure	One double and two single bonds	All single bonds	One double and two single bonds	NA
Electrical conductivity	Conductor	Insulator	Semiconductor to conductor	Semiconductor and metallic

Table 1.
Properties of different carbon allotropes.

2.2 Single-walled carbon nanotubes

Single-walled nanotubes (SWNTs) have a diameter of close to 1 nm, with a tube length that can be many thousands of times larger than their diameter. Single-walled nanotubes with length up to orders of centimeters have been produced. The structure of a SWNT may be envisaged by wrapping a one-atom-thick layer of carbon known as graphene into a seamless cylinder. The approach of wrapping the graphene sheet is drawn by a combination of indices (n,m) known as the chiral vector. The integers n and m denote the number of unit vectors on two directions within the honeycomb space lattice of graphene. Assuming that $m = 0$ with $\theta = 0°$, the nanotubes are called "zigzag." If n is equal to m with $\theta = 30°$, the nanotubes are called "armchair." Otherwise, they are known as "chiral" (when m and n are not the same) with $0 < \theta < 30°$ [8].

2.3 Multi-walled carbon nanotubes

Multi-walled nanotubes (MWNTs) can be considered as a gathering of concentric SWNTs with various diameters. The length and diameter of these structures contrast a great deal from those of SWNTs and, obviously, their properties are also very different. There are two models that can be utilized to depict the structures of MWNTs. In the event that the sheets of graphite are organized in concentric chambers (single-walled nanotube inside a bigger single-walled nanotube), then it is Russian doll model, whereas in the parchment model, a single sheet of graphite is rolled in around itself, resembling a scroll of parchment or a rolled up newspaper. The interlayer distance is close to the distance between graphene layers in graphite [9].

3. Different properties of CNTs

3.1 Electrical conductivity

CNTs can be very conducting and henceforth can be said to be metallic. Their conductivity has been demonstrated to be a function of their chirality, the level of bend just as their width. CNTs can be either metallic or semidirecting in their electrical conduct. Conductivity in MWNTs is quite complex. A few sorts of "arm chair"-structured CNTs seem to lead superior to other metallic CNTs. Moreover, the current between divider responses inside multi-walled nanotubes has been found to redistribute over individual cylinders nonconsistently. Be that as it may, there is no variation in current across various parts of metallic single-walled nanotubes. The conduct of the ropes of semiconducting single-walled nanotubes is unique; in that, the transport current changes unexpectedly at different situations on the CNTs.

3.2 Strength and elasticity

The carbon atoms of a solitary sheet of graphite structure a planar honeycomb cross segment, in which each atom is related by means of a strong bond to three neighboring atoms. With a perspective on these solid bonds, the basal plane flexible modulus of graphite is one of the greatest of any known material. In this way, CNTs are depended upon to be extremely high-quality fibers. Single-walled nanotubes are stiffer than steel and are incredibly impenetrable by damage from physical forces. Pushing on the tip of a nanotube will make it twist, yet without damage to the tip. Exactly when the force is cleared, the nanotube returns to its unique state. This property makes CNTs significant as test tips for high-resolution scanning probe microscopy. Assessing these effects has been to some degree inconvenient, and a cautious numerical worth has not been settled upon. Utilizing atomic force microscopy, the unanchored terminations of an unattached nanotube can be pushed out of its balance position, and the power required to push the nanotube can be assessed. The present Young's modulus estimation of single-walled nanotubes is around 1 Tpa. Others have shown theoretically that the Young's modulus depends upon the size and chirality of the single-walled nanotubes going from 1.22 to 1.26 Tpa.

3.3 Thermal conductivity and expansion

CNTs have been appeared to show superconductivity underneath 20°K (roughly at −253°C). Investigations propose that these intriguing strands, as of now proclaimed for their unparalleled strength and unique capacity to promote the electrical properties of either semiconductors or perfect metals, may sometime in the not-so-distant future too discover applications as smaller than expected warm conduits in gadgets and materials. The solid in-plane graphitic carbon–carbon bonds make them especially solid and solid against pivotal strains. The nearly zero in-plane thermal expansion but huge interplane extension of single-walled nanotubes infer solid in-plane coupling and high plasticity against nonaxial strains. Numerous applications of CNTs, such as in nanoscale molecular electronics, sensing and actuating gadgets, or as strengthening-added substance filaments in utilitarian composite materials, have been proposed. CNTs appear exceptionally as a tall warm conductivity. It is anticipated, hence, that nanotube reinforcement in polymeric materials may also appreciably progress the thermal and thermomechanical properties of the composites.

3.4 Field emission

Field emission comes about from the tunneling of electrons from a metal tip into vacuum, underneath utilization of a strong electric field. The little breadth and high aspect ratio of carbon nanotubes is extraordinarily ideal for field discharge. Undeniably for direct voltages, a strong electric field makes at the free conclusion of backed carbon nanotubes because of their sharpness.

DOI: http://dx.doi.org/10.5772/intechopen.90318

3.5 High aspect ratio

Carbon nanotubes address an extremely minimal high aspect ratio conductive-added substance for plastics of numerous types. Their high aspect ratio suggests that a lower stacking of carbon nanotubes is required contrasted with other conductive-added substances to perceive the equivalent electrical conductivity. This low loading protects more of the polymer resins' robustness, mostly at low temperatures, as well as keeping up other key performance properties of the matrix resin. Carbon nanotubes have shown to be an astounding added substance to give electrical conductivity in plastics. Their high angle proportion, around 1000:1, gives electrical conductivity at lower loadings, contrasted with standard included substance materials, for example, carbon dark, cleaved carbon fiber, or stainless steel fiber.

3.6 Highly absorbent

The huge surface area and high absorbency of CNTs make them perfect contender to utilize in air, various gases, and water purification. A lot of investigation is being done in substituting activated charcoal with CNTs in certain ultra high purification purposes.

4. Synthesis of carbon nanotubes

The development of carbon nanotubes during synthesis accepted to commence from the recombination of carbon molecules came apart by warm from its precursor. In spite of the fact that a number of more up-to-date fabrication procedures are being designed, three fundamental strategies are the laser ablation, electric arc discharge, and the chemical vapor deposition. Chemical vapor deposition is getting to be exceptionally well known because of its potential for scale-up generation (**Figure 1**).

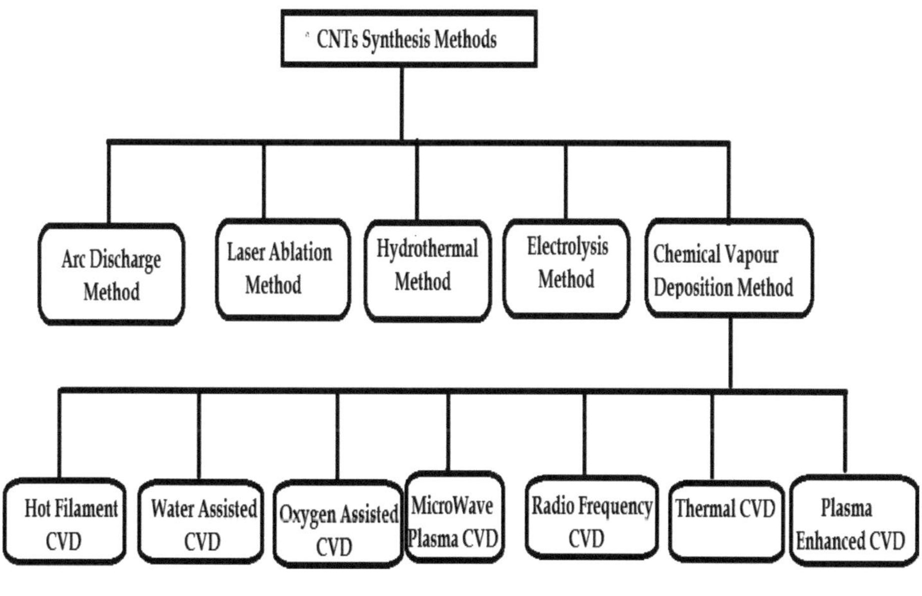

Figure 1.
Schematic diagram of different synthesis methods of CNTs.

4.1 Arc discharge

The arc discharge strategy produces a number of carbon nanostructures such as fullerenes, whisker, carbon soot, and profoundly graphitized carbon nanotubes from high-temperature plasma that approaches 3700°C. The crucial ever delivered nanotube was fabricated with the DC arc discharge methodology between two carbon anodes, anode and the cathode in a good gas (helium or argon) condition. For the most part, large-scale yield of carbon nanotubes of roughly 75% was made by Ebbesen and Ajayan with measurement between 2 and 30 nm and length 1 μm kept on the cathode at 100–500 torr helium and around 18 V DC. It has supportively been used to deliver both SWNTs and MWNTs as revealed by transmission electron microscope (TEM) assessment. Typical nanotube deposition rate is around 1 mm/min and the combination of transition metals, for example, cobalt, nickel, or iron into the terminals as impetus, favors nanotubes course of action against different nanoparticles, and low working temperature. The arc discharge unit must be given with cooling component whether catalyst is utilized or not, since overheating would not as it were comes about into safety dangers, but also into coalescence of the nanotube structure [10].

4.2 Laser ablation

Laser ablation strategy incorporates the use of laser bar to vaporize an object of a mix of graphite and metal impetus, for example, cobalt or nickel at temperature around 1200°C in a surge of controlled inert gas and weight, where the nanotube deposits are recuperated at a water cooled collector at much lower and steady temperature. This system was used in early days to deliver ropes of SWNTs with strikingly uniform limited width reaching out from 5 to 20 nm, and high return with graphite change more noticeable than 70–90% [11]. In any case, by the ethicalness of relative operational complexity, the laser removal strategy shows up to be financially disadvantageous, which on impact slow down its scale-up possibilities as compared to the CVD strategy.

4.3 Hydrothermal process

In a typical synthesis, ferrocene and sulfur in 1:2 proportion were dissolved in a mixed solution of water, and ethanol and NaOH pellets were added to the above solution under stirring for 30 min in a magnetic stirrer. The resulting homogeneous mixture was transferred to the reactor. The autoclave was heated to 200°C and maintained at this temperature for 20 h and then it was cooled to room temperature naturally and the resulting black precipitate was filtered, washed, and dried at 60°C in air [12].

4.4 Electolysis process

Soluble base and alkaline earth chloride salts are generally hygroscopic. Thermally drying (250°C) the salt in air can typically be acceptable for CNT production. (For CNT production, on account of the utilization of graphite crucible, melting the salt ought to be directed under inert climate). The tests were led at temperatures to some extent over the melting point of LiCl (600°C). To prepare the example approximately, the blend comprises of LiCl-1% $SnCl_2$ was taken in a graphite crucible. The graphite crucible was put in an electrical warmed heater and the temperature level set at 600°C. The free streaming inert gas was passed onto the furnace all through the experiment to diminish the oxidation of salt and carbon materials. The temperature came to 600°C and to equilibrate the melt for 1 h. This

Production method	Name of the product	Comments on product	Reaction conditions (catalyst)	Year & references	Author
Water assisted	Vertically aligned SWNTs, DWNTs	High yield	With buffer layer	2003 [15]	Liu et al.
O_2 assisted				2004 [16]	Chaisitsak et al.
Microwave plasma-enhanced				2004 [17]	Hata et al.
Hot filament-enhanced	SWNTs, MWNTs	Perpendicularly or vertically aligned	Fe-Co/SiO_2 with or without of Si support	2007 [18]	Zhu et al.
Alcohol CVD	CNTs	Multibranched morphology	Cu/MgO	2006 [19]	Terrado et al.
High power laser pulse alcohol CVD	SWNTs		Solid metal target	2005 [20]	Okamoto et al.
Alcohol CVD		High purity	Ferrocene-ethanol	2002 [21]	Maruyama et al.
Thermal CVD	CNTs	Aligned	Co/SiO_2, Ar/H_2 and NH_3/N_2	2009 [22]	Khatri et al.
Microwave plasma-enhanced	CNTs	Well aligned, curved with random orientation	Fe/sapphire, Ni-Fe/glass, Cr-Fe/glass, Fe/Si, stainless steal	2009 [23]	Qi et al.
CVD	CNTs, carbon onions		Ni/Al	2006 [24]	Zhao et al.
		Metal filled, bamboo shaped	Ni/Cu/Al, methane	2008 [25]	Kang et al.
	MWNTs		K-doped Co and Co-Fe/zeolite and $CaCO_3$	2008 [26]	Balogh et al.
	SWNTs, DWNTs		Fe-Mo/MgO	2006 [27]	Ago et al.
	CNTs		Different metals and rare-earth promoters	2000 [28]	Willems et al.
CVD	Aligned CNTs		Single-crystal of sapphire or quartz	2004 [29]	Ismach et al.
	CNTs, graphite layers, filaments		Different types of catalysts	1999 [30]	Fan et al.
	Helicoidal CNTs	Regular and irregular shape		1995 [31]	Ihara et al.
Alcohol CCVD	CNTs	Various morphology depending on the metal film thickness	Co/Si, Co-Mo/Si, Co/quartz, Co-Mo/quartz	2004 [32]	Nishide et al.
Thermal CVD	SWNTs & MWNTs	Controlled diameter distribution	Fe-Co/Zeolite	2005 [33]	Mukul Kumar et al.

Production method	Name of the product	Comments on product	Reaction conditions (catalyst)	Year & references	Author
CVD	MWNTs	Pure & high quality	$CaCO_3$	2003 [34]	Herandi et al.
CVD	MWNTs	Clear helical structure	Iron nanoparticle catalyst	1993 [35]	Yacaman et al.

Table 2.
Carbon nanotubes synthesized by CVD method using different carbon sources.

Production method	Name of the product	Comments on product	Reaction conditions (catalyst)	Year and references	Author
Spray pyrolysis	MWNTs	Bamboo-shaped structure	*Brassica juncea*	2014 [37]	Kalaiselvan et al.
		Well-graphitized MWNTs	*Madhuca longifolia*	2013 [38]	
		Well-graphitized MWNTs	Pine oil	2010 [39]	Karthikeyan et al.
		Well-graphitized MWNTs	Bio-diesel oil	2010 [40]	
			Cymbopogon flexuosus oil	2012 [41]	Mageswari et al.
		Vertically aligned CNTs	*Helianthus annuus* oil	2013 [42]	Angulakshmi et al.
		Entangled MWCNTs	*Madhuca longifolia* oil	2014 [43]	Kalaiselvan et al.
			Cymbopogon flexuosus oil	2014 [44]	Mageswari et al.
	MWNTs	**Magnetic metal encapsulated**	*Pongamia pinnata* oil	2014 [45]	Mahalingam et al.
	Few-walled CNTs		Aliphatic alcohols	2013 [46]	Ordonez Casanova et al.
	CNTs	**Nitrogen doped**	Imidazole and Acetonitrile	2011 [47]	Jian Liu et al.
	MWNTs	Well-graphitized MWNTs	*Oryza sativa* oil	2016 [48]	Kalaiselvan et al.
		Multilayer of walls	*Citrus limonum* oil	2018 [49]	Angulakshmi et al.
		Multilayer of walls	*Zingiber officinale* oil	2018 [50]	Kalaiselvan et al.

Table 3.
Carbon nanotubes synthesized by spray pyrolysis method using different carbon sources.

procedure was called pre-electrolysis. After the pre-electrolysis, graphite bar of 1.4 cm^2 diameter with 5 cm^2 height was embedded into the melt as cathode. The cathode was embedded into the melt up to 4 cm^2 deepness in the graphite crucible as anode just as holding vessel for electrolyte. The control of DC power supply additionally needs to give an adequately wide range of currents and voltages. This procedure was completed at the steady present 3 A, and voltage 4.6 V. The current, voltage, and temperature of the bath are recorded all through the procedure [13].

4.5 Chemical vapor deposition

In this system, carbon nanotubes were developed from the breakdown of hydrocarbons at temperature range of 500–1200°C. They can develop on substrates for example, carbon, quartz, and silicon or on floating fine catalyst particle like Fe, Ni, Co, and so forth from various hydrocarbons for example, benzene, xylene, flammable gas, and acetylene. CVD furnished with a level cylindrical heater as the reactor. The cylinder is made of quartz and is 30 mm in width and 1000 mm long. Ferrocene and benzene vapors go about as the catalyst (Fe) and carbon atom precursor separately were moved by argon, hydrogen, or blend of both into the response chamber, and disintegrated into individual particles of Fe and carbon atom, coming about into carbon nanostructures. The development of the nanostructures happened in the heating zone, previously, or after the heating zone, which has, on the whole, functioned somewhere in the range of 500 and 1150°C for around 30 min. About 200 ml/min of hydrogen is utilized to cool the reactor [14] (**Tables 2** and **3**).

4.6 Spray pyrolysis

Progress metal nanoparticles were directly put in a quartz boat and kept at the center of a quartz tube, which was put inside a tubular heater. The inert gas nitrogen was presented at a rate of 100 mL/min into the quartz tube to expel any oxygen from that point and to make inert environment (**Figures 2** and **3**). The temperature was raised from room temperature to the preferred growth temperature. In this way, carbon precursor was brought into the quartz tube through a spray nozzle and the stream was kept up at a rate of 0.5 mL/min at temperature of 650°C. This process was carried out for 45 min and thereafter the furnace was cooled to room temperature. Nitrogen atmosphere was maintained throughout the experiment [36].

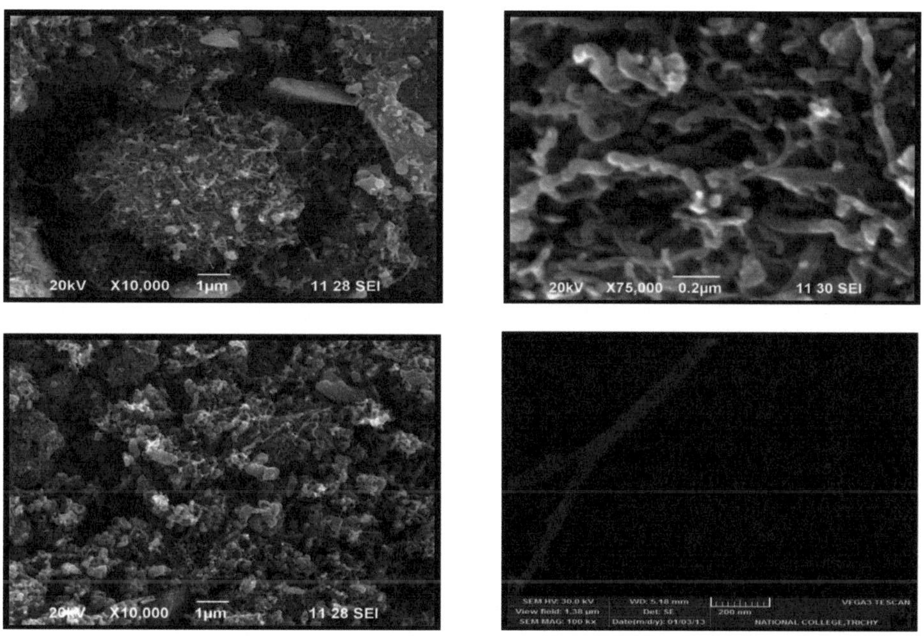

Figure 2.
SEM images of carbon nanotubes from different carbon sources.

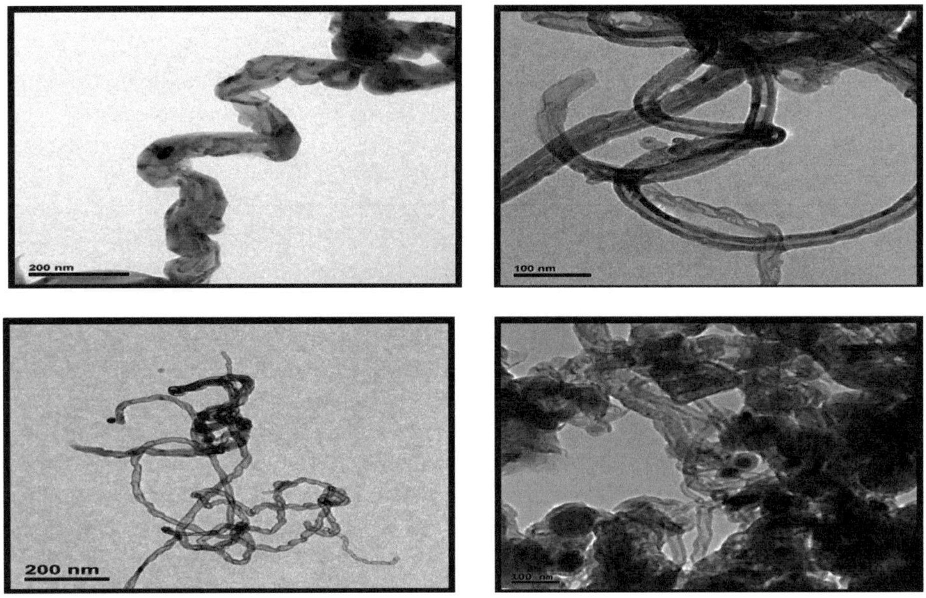

Figure 3.
HRTEM images of carbon nanotubes from different carbon sources.

5. CNT growth mechanism

CNT growth mechanism has been debatable right from its discovery. Quite a lot of groups have proposed a few feasible outcomes that are often intricate. Subsequently, no single CNT growth pattern is well recognized till date. Despite the fact that most broadly acknowledged general system can be laid out as follows, hydrocarbon vapor when interacting with the "hot" metal nanoparticles first decompose into carbon and hydrogen species; hydrogen takes off and carbon gets broken down into the metal. Subsequent to achieving the carbon solvency limit in the metal at that temperature, as-disintegrated carbon accelerates out and takes shape as a tube-shaped system having no dangling bonds and hence actively stable.

Presently, there are two general cases. At the point when the impetus substrate collaboration is powerless, hydrocarbon decays on the top surface of the metal, carbon diffuses down through the metal, and CNT encourages out over the metallic base, pushing the whole metal molecule off the substrate insofar as the metal's top is open for sparkling hydrocarbon decomposition and CNT continues to grow longer and more when the metal is completely encased with surplus carbon, its reactant movement arrives to an end, and the CNT development is halted up. This sort of growth is identified as tip-growth mechanism. In the other case, when the catalyst substrate interaction is strong, starting hydrocarbon decay and carbon dispersion occur like that in the tip-growth case; however, the CNT precipitation fails to drive the metal particle up so the precipitation is bound to rise out from the metal's top. To begin with, carbon crystallizes out as a hemispherical ring, which at that point stretches out up as seamless graphitic cylinder. Consequently, hydrocarbon deposition happens on the lower fringe surface of the metal, and as broken down carbon diffuses upward. Accordingly, CNT grows up with the catalyst molecule established on its base; henceforth, this is considered as "base-development model." The formation of single- or multi-walled CNT is governed by the size of the catalyst nanoparticle. Generally speaking, when the particle size is a few nm, SWCNT forms, whereas if the particles are a few tens of nm wide, it favors MWCNT growth [51] (**Figure 4**).

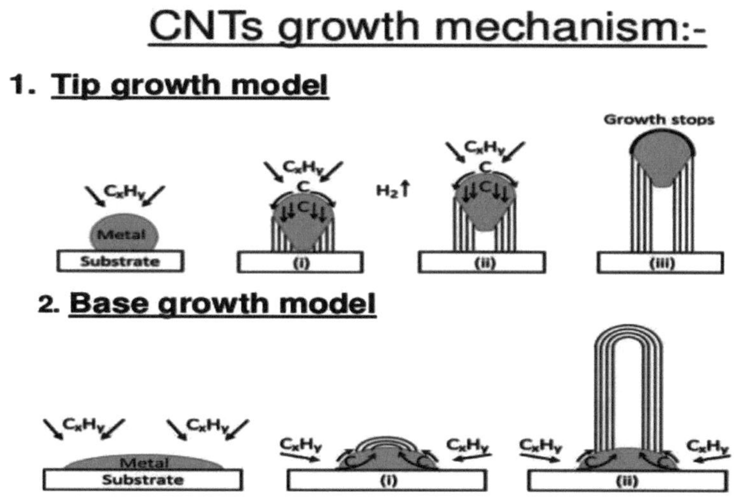

Figure 4.
Schematic diagram of growth mechanisms of carbon nanotubes.

6. Purification process

Many purification methods that are highly developed show the most promise for industrial-scale production in which most type of impurities such as graphitic nanoparticles, amorphous carbon, fullerenes particles containing (transition) metal catalyst support removed since as-produced CNT soot contains a lot of impurities. These impurities will interfere with most of the desired properties of the CNTs. The common industrial techniques use strong oxidation and acid refluxing techniques, which have an effect on the structure of the tubes. In this chapter, several purification techniques of the CNT will be discussed such as air oxidation, acid treatment, annealing, ultrasonication, microfiltration, ferromagnetic separation, cutting, functionalization, and chromatography techniques. Most of the techniques used are combined with other techniques.

6.1 Air oxidation

Air oxidation treatment of the CNTs is a useful way to eliminate carbon impurities (defect carbon) and metal catalyst. The main disadvantage of this process is that not only the impurities are oxidized but also the CNTs. The reason why impurity oxidation is preferred is that these impurities are most commonly attached to the metal catalyst, which also acts as oxidizing catalyst. Optimum yield of this process is highly dependent on a lot of factors, such as metal content, oxidation time, oxidizing agent, and temperature. When the temperature is raised above 550°C, CNTs will also be oxidized. Optimum oxidation condition is found to be 400°C for 40 min [52].

6.2 Acid treatment

Commonly, the acid treatment will evacuate the metal catalyst. Most importantly, the outside of the metal must be exposed by oxidation or ultra-sonication. The metal catalyst is then presented to be destructive and solvated. The CNTs stay in suspended structure. When utilizing a treatment in HNO_3, the corrosive just affects the metal catalyst. It has no impact on the CNTs and other carbon particles. On the other hand, if a treatment in HCl is utilized, the corrosive has additionally a little impact on the CNTs and other carbon particles [53, 54].

6.3 Annealing

On account of high temperatures (500–1500°C), the nanotubes will be redesigned and imperfections will be removed [55]. The high temperature furthermore causes the graphitic carbon and the short fullerenes to pyrolysis. When using high temperature vacuum treatment (1500°C), the metal will be mollified and can moreover be evacuated [54].

6.4 Ultrasonication

This process involves that nanoparticles be isolated utilizing ultrasonic vibrations. Agglomeration of various nanoparticles will be compelled to vibrate and will turn out to be scattered. The detachment of the particles is profoundly relying upon the surfactant, dissolvable, and reagent utilized. The stability of the scattered tubes is impacted by the solvent in the system. In poor solvents, the CNTs are increasingly secure in the event that they are as yet attached to the metal. In any case, in certain solvents like alcohols, even monodispersed particles are generally steady. At the point when acid is utilized, the wholesomeness of the CNTs relies upon the contact time. At the point when the tubes are bare to the acid for a short time frame, just the metal solvates, yet for a more drawn out for exposure time, the cylinders will also be chemically reduced [56, 57].

6.5 Magnetic purification

In this process, ferromagnetic nanoparticles are mechanically separated from their graphitic lattice [58]. The CNTs' suspension is blended with inorganic nanoparticles of ZrO_2 or $CaCO_3$ in a ultrasonic bath to take away the ferromagnetic particles. At that point, the particles are bound with permanent magnetic poles. After a successive chemical treatment, a highly cleaned CNT material will be obtained. This procedure does not require the involvement of big equipment and facilitates the fabrication of optimal quantities of CNTs free of magnetic impurities.

6.6 Microfiltration

This method is employed to remove carbon nanoparticles by microfiltration, which is completely based on size or particle separation. The other nanoparticles such as catalyst metal and fullerenes are also passing through the filter. One way of separating fullerenes from the CNTs by microfiltration is to immerse the as-synthesized CNTs first in a carbon-di-sulfide solution. The contents that are insoluble in carbon-di-sulfide (as-synthesized CNTs) are then trapped in a filter. The fullerenes that are solvated in the carbon-di-sulfide pass through the filter [57].

6.7 Cutting

Cutting of the CNTs is done either by chemically induced method or by mechanical cutting or as a combination of these two. Mechanical cutting of the nanotubes can be induced by ball-milling. Here, the bonds will break due to the high friction between the nanoparticles and the nanotubes will be disordered. CNTs can be chemically cut by partial function of the tube structures, for instance, with fluorated carbon. Then, the fluorated carbon will be driven off the sidewall with pyrolyzation as CF_4 or COF_2. This will discard the chemically hacked nanotubes. A blend of mechanical and chemical cutting of the nanotubes is ultrasonical empowered cutting in acid solution treatment. Along this way, the ultrasonic vibration will give the nanotubes adequate energy to leave the catalyst surface. At that point, while blending with acid, the nanotubes will break at the deformity locales [59].

6.8 Functionalization of carbon nanotubes

Functionalization depends on making carbon nanotubes hydrophilic in nature than the contaminations by appending different groups to the tubes. Presently, it is anything but difficult to isolate them from insoluble contaminations, for example, metal particles by filtration. An added functionalization procedure also leaves the carbon nanotubes' structure unblemished and makes them solvent for chromatographic size division. For recovery of the purified carbon nanotubes, the helpful functional moieties can be evidently cleared by heat treatment process, such as annealing [60].

6.9 Chromatography

This system is primarily used to isolate little amounts of CNTs into fractions with little length and diameter dispersion. The CNTs overflow into a column comprised of permeable material, through which the CNTs will run. The columns utilized are gel permeation chromatography and high performance liquid chromatography-size exclusion chromatography columns. The number of pores the CNTs will move through relies upon their size. This implies, the littler the particle, the more drawn out the pathway as far as possible of the section will be and that the bigger atoms will succeed first. However, a problem is that the CNTs have to be either dispersed or solvated. This can be done by ultrasonication or functionalization with soluble groups [61].

7. Characterization of CNTs

Thermogravimetric analysis, scanning electron microscopy, transmission electron microscopy, atomic force microscopy, Raman spectroscopy, infrared spectroscopy, and nuclear magnetic resonance have been used. While TEM, SEM, and AFM have been used for the majority part to qualitatively ascertain the widespread structural studies of carbon nanotubes, infrared spectroscopy, Raman spectroscopy, and nuclear magnetic resonance spectroscopy have been used to authenticate the presence of useful moiety on carbon nanotubes. All procedures have advantages when used in combination with other techniques [62].

7.1 Transmission electron microscopy

It is acclimated affirm the morphology and to give quantitative insight into the purity of incorporated carbon nanotubes. TEM unambiguously gives subjective information on the size, shape, and structure of carbonous materials, in addition to non-CNT–structured contaminations in the sample. Be that as it may, it cannot spot metallic pollutions and does not separate from CNTs. TEM has furthermore been acquainted image cell take-up of CNT-drug composites and to see the result of the CNT component once cell take-up [63].

7.2 Scanning electron microscopy

This is one of the main surface analysis instruments. It is used in the preliminary evaluation of CNT morphology. It can measure the diameter of CNT not precisely but roughly. In its conventional situation, the technique is controlled by its incapacity to set apart catalyst and carboniferous impurities from CNTs. However, the metallic content of CNT samples is customarily quantifiable by SEM tied with an energy dispersive X-ray analysis detector (SEM-EDX). Despite the consequences, SEM is perhaps the only technique that will give data on each CNT structural information and also the metal impurity content [64].

7.3 Raman spectroscopy

Raman spectroscopy is one of the useful methods to detect carbon nanotubes, which not only shows the regularity and purity of the sample but also defines the diameter distribution of carbon nanotubes. Raman spectroscopy is used with a laser excitation wavelength of 633 nm. In the Raman spectra, there are three peaks or regions we are concerned about: the radial breathing modes (RBM ~ 100–300 cm^{-1}), D peak (~1350 cm^{-1}), and G peak (~1570 cm^{-1}). The RBM peaks are the distinctive peaks of carbon nanotubes, analogous with the diameters of carbon nanotubes. From radial breathing mode, we can forecast the distribution of carbon nanotubes' diameters. The D and G peaks are a result of unstructured carbon and graphited carbon, correspondingly. We can approximate the purity of carbon nanotubes by the intensity ratio of G peak and D peak (G/D). The larger the resultant value of G/D, the higher the graphited carbon, and if there are not many impurities or defects, then the purity is higher [65].

7.4 Proton nuclear magnetic resonance

It has been accustomed to monitor the advancement of CNT functionalization. The proximity of useful moiety can be predicted by characteristic peaks emerging from the differentiation within the magnetic environment. H^1-NMR of functionalized CNTs is portrayed by wide peaks for protons close to the CNT, fitting sharper with separation. H^1-NMR has been accustomed to analyze the synthesis and attachment of functional moiety to CNTs [66].

7.5 Infrared spectroscopy

It is for the most part a subjective method used to recognize valuable moieties and the idea of their connection to CNT sidewalls. Characteristic useful moieties incorporate characteristic frequencies of IR radiation, giving rise to a fingerprint identification of bonds. It is a contrary system to NMR, to affirm the presence of bonds among CNTs and of joined moieties [66].

7.6 Thermo gravimetric analysis (TGA)

It is fundamentally a subjective instrument acclimated set up useful gatherings and furthermore the idea of their connection to CNT sidewalls. Very surprising functional moieties retain characteristic frequencies of IR radiation, offering prompt fingerprint recognition of bonds. It is a harmonizing technique to magnetic resonance, to substantiate the presence of bonds between CNTs and of coupled moieties. The high sensitivity of the TGA, which is in the order of 0.1 mg/min, acceptable weight loss determinations at a given heating rate, within a short time without overwhelming too much material. The oxidation rates of carbon nanotubes measured in air at atmospheric pressure within the TGA are exceptional for each CNT sample of diverse wall thicknesses [67].

8. Recent applications of CNTs

From the time when the discovery of carbon nanotubes was made in 1991, several research teams reported their potential applications in diverse fields including energy storage, molecular electronics, nanoprobes, nanosensors, nanotube composites, and nanotube templates based on their unique electronic properties, size, mechanical strength, and flexibility (**Table 4**).

Potential applications	Author and year	Ref
Medicine applications		
Solid phase extraction of drugs and biochemicals	Xiao et al. 2013	[68]
Drug delivery for cancer therapy	Zhang et al. 2011	[69]
Antioxidants	Pham-Huy et al. 2008	[70]
Antitumor immunotherapy	Yang et al. 2007	[71]
Local antitumor hyperthermia therapy	Madani et al. 2011	[72]
Infection therapy	Jiang et al. 2012	[73]
Gene therapy by DNA delivery	Liao et al. 2011	[74]
Tissue regeneration	MacDonald et al. 2005	[75]
Artificial implants	Zhang et al. 2010	[76]
Biosensor vehicles for diagnostics and detection	Wang 2005	[77]
Neurodegenerative diseases	Zhang et al. 2010	[76]
Alzheimer syndrome	Yang et al. 2010	[78]
Other potential applications		
Composite materials	Zhou et al. 2014	[79]
Coatings and films	Mirri et al. 2012	[80]
Microelectronics	Duesberg et al. 2003	[81]
Transistors	Aissa et al. 2015	[82]
Solar cells	Wang et al. 2015	[83]
Hydrogen storage	Adeniran et al. 2015	[84]
Interconnects	Li et al. 2013	[85]
Electronic components	Cai et al. 2015	[86]
Loudspeaker	Xiao et al. 2008	[87]
Environment	Ong et al. 2010	[88]
Biosensors	Xia et al. 2015	[89]
Superhydrophobic cotton fabric	Makowski et al. 2014	[90]
Oscillators	Kang et al. 2006	[91]
Light bulb filament	Wei et al. 2004	[92]
Magnets	Kyatskaya et al. 2009	[93]
Electromagnetic antenna	Maksimenko et al. 2008	[94]
Air pollution filter	Yildiz et al. 2013	[95]
Water filter	Das et al. 2014	[96]
Nanowires for light-emitting devices	Yu et al. 2015	[97]

Table 4.
Summary of CNTs-based applications.

9. Conclusion

In this work, we have summarized various synthesis techniques of carbon nanotubes, and this advanced carbon nanomaterial has emerged since their discovery about 25 years ago. The broad scope of these applications suggests that a large number of CNT-based technologies will result from their unique properties of high electrical conductivity, mechanical strength, high aspect ratio, and nanoscale diameter. This report on carbon nanotubes reveals also an overview on structure, morphology, purification, and characterization methods. The distinct structural properties of carbon nanotubes, in particular their high aspect ratio, strength, and high surface area, have the added advantage of being potential nanodevices for various medical and industrial applications. Overall, recent studies regarding CNTs have shown a very promising glimpse of what lies ahead in the future of science. In this work, we also reported a variety of familiar methods to synthesis carbon nanotubes such as arc discharge, laser ablation and spray pyrolysis-assisted chemical vapor deposition, types, properties, purifications, characterization, and some notable applications in diverse field and also covered some recent synthesis of CNTs from a range of hydrocarbons using spray pyrolysis.

Acknowledgements

The authors acknowledge the management of SNS College of Technology, Coimbatore, India, for their constant encouragement and support.

Author details

Kalaiselvan Shanmugam*, J. Manivannan and M. Manjuladevi
Department of Chemistry, SNS College of Technology, Coimbatore, Tamilnadu, India

*Address all correspondence to: kalaichem82@gmail.com

References

[1] Kroto HW, Heath JR, O'Brian SC, Curl RF, Smalley RE. Nature. 1985; **381**:162

[2] Kratschmer W, Lamb LD, Fostiropoulos K, Huffman DR. Nature. 1990;**347**:354

[3] Journet C, Maser WK, Bernier P, Loiseau A, Lamy de la Chapelle M, Lefrant S, et al. Large-scale production of single-walled carbon nanotubes by the electric-arc technique. Nature. 1997;**388**:756-758

[4] Iijima S, Ichihashi T. Single-shell carbon nanotubes of 1-nm diameter. Nature. 1993;**363**:603-605

[5] Alford JM, Mason GR, Feikema DA. Formation of carbon nanotubes in a microgravity environment. In: Sixth International Microgravity Combustion Workshop. NASA Glenn Research Center, Cleveland, OH, CP-2001-210826, May 22-24. 2001. pp. 293-296

[6] Yamabe T. Recent development of carbon nanotubes. Synthetic Metals. 1995:1511-1518

[7] Marsh H, Rodríguez-Reinoso F. Activated Carbon. Elsevier; 2006

[8] Zhu J, Kim J, Peng H, Margrave JL, Khabashesku VN, Barrera EV. Improving the dispersion and integration of single-walled carbon nanotubes in epoxy composites through functionalization. Nano Letters. 2003;**3**:1107-1113

[9] Bianco A, Kostarelos K, Prato M.Making carbon nanotubes biocompatible and biodegradable. ChemCommun (Camb). 2011;**47**:10182-10188

[10] Iijima S. Helical microtubules of graphitic carbon. Nature. 1991;**354**:56-58

[11] Guo T, Nikolaev P, Thess A, Colbert DT, Smalley RE. Catalytic growth of single-walled nanotubes by laser vaporization. Chemical Physics Letters. 1995;**243**(1-2):49-54

[12] Krishnamurthy G, Namitha R. A novel method of synthesis of carbon nanotube by hydrothermal process. International Journal of Science Research. **1**(4):358-362

[13] Anagappana S, Thirumal V, Ramkumar K, Visuvasam A. Synthesis of carbon nanotubes by molten salt technique. Chemical Science Transactions. 2013;**2**(2):575-583. DOI: 10.7598/cst2013.394

[14] Li WZ, Xie SS, Qian LX, Chang BH, Zou BS, Zhou WY, et al. Large scale synthesis of aligned carbon nanotubes. Science. 1996;**274**(5293):1701-1703. DOI: 10.1126/science.274.5293.1701

[15] Liu GY, Zhong DY, Xia SH, Cheng SF, Ding YG, Lu YJ, et al. CNTs grown on the surface of various materials by large volume MP-CVD for VME applications. Applied Surface Science. 2003;**215**:209-213

[16] Chaisitsak S, Yamada A, Konagi M. Hot filament enhanced CVD synthesis of carbon nanotubes by using a carbon filament. Diamond Rel. Mat. 2004;**13**:438-444

[17] Hata K, Futaba DH, Nizuno K, Namai T, Yumura M, Iijima S. Water-assisted highly efficient synthesis of impurity-free single-walled carbon nanotubes. Science. 2004;**306**:1362-1364

[18] Zhu M, Wang J, Outlaw RA, Hou K, Manos DM, Holloway BC. Synthesis of carbon nanosheets and carbon nanotubes by radio frequency plasma enhanced chemical vapor deposition. Diamond Rel. Mat. 2007;**16**:196-201

[19] Terrado E, Redrado M, Munoz E, Maser WK, Benito AM, Martinez MT. Carbon nanotube growth on cobalt-sprayed substrates by thermal CVD. Materials Science and Engineering C. 2006;**26**:1185-1188

[20] Okamoto A, Shinohara H. Control of diameter distribution of single-walled carbon nanotubes using the zeolite-CCVD method at atmospheric pressure. Carbon. 2005;**43**:431-436

[21] Maruyama S, Kojima R, Miyauchi R, Chiashi S, Kohno M. Low-temperature synthesis of high-purity single-walled carbon nanotubes from alcohol. Chemical Physics Letters. 2002;**360**:229-234

[22] Khatri I, Soga T, Jimbo T, Adhikari A, Aryal HR, Umeno M. Synthesis of single walled carbon nanotubes by ultrasonic spray pyrolysis mehod. Diamond Related Materials. 2009;**18**:319-323

[23] Qi JL, Wang X, Tian HW, Peng YS, Liu C, Zheng WT. Syntheses of carbon nanomaterials by radio frequency plasma enhanced chemical vapor deposition. Journal of Alloys Compounds. 2009;**486**:265-272

[24] Zhao NQ, He CN, Du XW, Shi CS, Li JJ, Cui L. Amorphous carbon nanotubes fabricated by low-temperature chemical vapor deposition. Carbon. 1859-1862;**2006**:44

[25] Kang J, Li J, Du X, Shi C, Zhao N, Nash P. Synthesis of carbon nanotubes and carbon onions by CVD using a Ni/Y catalyst supported on copper. Materials Science and Engineering A. 2008;**475**:136-140

[26] Balogh Z, Halasi G, Korbely B, Hernadi K. CVD synthesis of MWNTs over K-doped supported catalysts. Applied Catalysis A: General. 2008;**344**:191-197

[27] Ago H, Uehara N, Yoshihara N, Tsuji M, Yumura M, Tomonaga N, et al. Gas analysis of the CVD process for high yield growth of carbon nanotubes over metal-supported catalysts. Carbon. 2006;**44**:2912-2918

[28] Willems I, Kónya Z, Colomer J-F, Van Tendeloo G, Nagaraju N, Fonseca A, et al. Control of the outer diameter of thin carbon nanotubes synthesized by catalytic decomposition of hydrocarbons. Chemical Physics Letters. 2000;**317**:71-76

[29] Ismach A, Segev L, Wachtel E, Joselevich E. Atomic-step-templated formation of single wall carbon nanotube patterns. Angewandte Chemie, International Edition. 2004;**43**:6140-6143

[30] Fan SS, Chapline MG, Franklin NR, Tombler TW, Cassell AM, Dai HJ. Self-oriented regular arrays of carbon nanotubes and their field emission properties. Science. 1999;**283**:512-514

[31] Ihara S, Itoh S. Helically coiled and toroidal cage forms of graphitic carbon. Carbon. 1995;**33**:931-939

[32] Nishide D, Kataura H, Suzuki S, Okubo S, Achiba Y. Growth of single-wall carbon nanotubes from ethanol vapor on cobalt particles produced by pulsed laser vaporization. Chemical Physics Letters. 2004;**392**:309-313

[33] Kumar M, Ando Y. Contolling the diameter distribution of carbon nanotubes grown from camphor on a zeolite support. Carbon. 2005;**43**:533-540

[34] Couteau E, Herandi K, Seo JW, Thien-Nga L, Miko C, Gaal R, et al. CVD synthesis of high-purity multiwalled carbon nanotubes using $CaCO_3$ catalyst support for large-scale production. Chemical Physics Letters. 2003;**378**:9-17

[35] Jose-Yacaman M, Miki-Yoshida M, Rendon L, Santiesteban JG. Applied Physics Letters. 1993;**62**:697

[36] Aguilar-Elguezabal A, Antunez W, Alonso G, Delgado FP, Espinosa F, Miki-Yoshida M. Study of carbon nanotubes synthesis by spray pyrolysis and model of growth. Diamond and Related Materials. 2006;**15**(9):1329-1335

[37] Kalaiselvan S, Karthik M, Vladimir R, Karthikeyan S. Growth of bamboo like carbon nanotubes from *Brassica Juncea* as natural precursor. Journal of Environmental Nanotechnology. 2014;**3**:92-100

[38] Karthikeyan S, Kalaiselvan S, Manorangitham D, Maragathamani S. Morphology and structural studies of multi-walled carbon nanotubes by spray pyrolysis using *Madhuca Longifolia* oil. Journal of Environmental Nanotechnology. 2013;**2**:15-20

[39] Karthikeyan S, Mahalingam P. Studies of yield and nature of multi-walled carbon nanotubes synthesized by spray pyrolysis of pine oil at different temperatures. International Journal of Nanotechnology and Applications. 2010;**4**:189-197

[40] Karthikeyan S, Mahalingam P. Synthesis and characterization of multi-walled carbon nanotubes from biodiesel oil: Green nanotechnology route. International Journal of Green Nanotechnology: Physics and Chemistry. 2010;**2**:39-46

[41] Mageswari S, Jafar Ahamed A, Karthikeyan S. Effect of temperature and flow rate on the yield of multiwalled carbon nanotubes by spray pyrolysis using *Cymbopogen flexsuous* oil. Journal of Environmental Nanotechnology. 2012;**1**:28-31

[42] Angulakshmi VS, Rajasekar K, Sathiskumar C, Karthikeyan S. Growth of vertically aligned carbon nanotubes by spray pyrolysis using green precursor—methyl ester of *Helianthus annuss* oil. New Carbon Materials. 2013;**28**:284-288

[43] Kalaiselvan S, Anitha K, Shanthi P, Syed Shabudeen PS, Karthikeyan S. Morphology of entangled multiwalled carbon nanotubes by catalytic spray pyrolysis using *Madhuca longifolia* oil as a precursor. Rasayan Journal of Chemistry. 2014;**7**(4):333-339

[44] Mageswari S, Kalaiselvan S, Syed Shabudeen PS, Sivakumar N, Karthikeyan S. Optimization of growth temperature of multi-walled carbon nanotubes fabricated by chemical vapour deposition and their application for arsenic removal. International Journal of Material Science Poland. 2014;**32**(4):709-718

[45] Mahalingam P, Sivakumar N, KarthikM,KarthikeyanS.Characterization of magnetic metal encapsulated in multi-walled carbon nanotubes synthesized from methyl ester of *Pongamia pinnata* oil and its application for removal of arsenic ions from aqueous solution. Asian Journal of Chemistry. 2014;**26**(14):4167-4171

[46] Ordonez-Casanova EG, Roman-AguirreM,Aguilar-ElguezabalA, Espinosa-Magana F. Synthesis of carbon nanotubes of few walls using aliphatic alcohols as a carbon source. Materials. 2013;**6**:2534-2542

[47] Liu J, Zhang Y, Ionescu MI, Li R, Sun X. Nitrogen-doped carbon nanotubes with tunable structure and high yield produced by ultrasonic spray pyrolysis. Applied Surface Science. 2011;**257**:7837-7844

[48] Kalaiselvan S, Balachandran K, Karthikeyan S, Venckatesh R. Renewable precursor based MWCNTs synthesized by spray pyrolysis method for DSSC application'. Silicon-Springer.

2016;**10**(2):211-217. DOI: 10.1007/s12633-016-9419

[49] Angulakshmi VS, Mageswari S, KalaiselvanS, KarthikeyanS. Application of Box Behnken design to optimize the reaction conditions on the synthesis of multiwalled carbon nanotubes. Journal of Environmental Nanotechnology. 2018;**7**(1):30-36

[50] Kalaiselvan S, Mathan Kumar N, Manivannan J. Production of multilayered nanostructure from *Zingiber officinale* by spray pyrolysis method. Global Journal of Science Frontier Research: B Chemistry. 2018;**18**(03)

[51] Coville NJ, Mhlanga SD, Nxumalo EN, Shaikjee A. A review of shaped carbon nanomaterials. South African Journal of Science. 2011;**107**(3/4). Art. #418, 15 pages. DOI: 10.4102/sajs.v107i3/4.418

[52] Li C, Wang D, Liang T, Wang X, Wu J, Hu X, et al. Oxidation of multiwalled carbon nanotubes by air: Benefits for electric double layer capacitors. Powder Technology. 2004;**142**:175-179

[53] Ka F, Lin Y, Sun YP. High aqueous solubility of functionalized single-walled carbon nanotubes. Langmuir. 2004;**20**:4777-4778

[54] Kajiura H, Tsutsui S, Huang HJ, Murakami Y. Chemical Physics Letters. 2002;**364**:586

[55] Borowiak-Palen E, Pichler T, Liu X, Knupfer M, Graff A, Jost O, et al. Chemical Physics Letters. 2002;**363**(5-6):567-572

[56] Shelimov KB, Esenaliev RO, Rinzler AG, Huffman CB, Smalley RE. Chemical Physics Letters. 1998;**282**:429

[57] Bandow S, Rao AM, Williams KA, Thess A, Smalley RE, Eklund PC. The Journal of Physical Chemistry. B. 1997;**101**:44

[58] Thien-Nga L, Hernadi K, Ljubovic E, Garaj S, Forro L. Nano Letters. 2002;**2**(12):1349-1352

[59] Liu Y, Gao L, Sun J, Zheng S, Jiang L, Wang Y, et al. A multi-step strategy for cutting and purification of single-walled carbon nanotubes. Carbon. 2007;**45**(10):1972-1978

[60] Aviles F, Cauich-Rodriguez JV, Moo-TahL,May-PatA,Vargas-CoronadoR. Evaluation of mild acid oxidation treatmentsforMWCNTfunctionalization. Carbon. 2009;**47**(13):2970-2975

[61] Hou P-X, Liu C, Cheng H-M. Purification of carbon nanotubes. Carbon. 2008;**46**:2003-2025

[62] Itkis ME, Perea DE, Jung R, Niyogi S, Haddon RC. Comparison of analytical techniques for purity evaluation of single-walled carbon nanotubes. Journal of the American Chemical Society. 2005;**127**:3439-3448

[63] Kim BM, Qian S, Bau HH. Filling carbon nanotubes with particles. Nano Letters. 2005;**5**:873-878

[64] Abdulkareem AS, Afolabi AS, Iyuke SE, Vz Pienaar HC. Synthesis of carbon nanotubes by swirled floating catalyst chemical vapour deposition method. Journal of Nanoscience and Nanotechnology. 2007;**7**:3233-3238

[65] Dresselhaus MS et al. Raman spectroscoy of carbon nantubes in 1997 and 2007. Journal of Physical Chemistry C. 2007;**111**(48):17887-17893

[66] Yinghuai Z, Peng AT, Carpenter K, Maguire JA, Hosmane NS, Takagaki M. Substituted carborane-appended water-soluble single wall carbon nanotubes: New approach to boron neutron capture therapy drug

delivery. Journal of the American Chemical Society. 2005;**127**:9875-9880

[67] Pang LSK, Saxby JD, Chatfield SP. Thermogravimetric analysis of carbon nanotubes and nanoparticles. Journal of Physical Chemistry. 1993;**97**:27

[68] Xiao D, Dramou P, Xiong N, et al. Development of novel molecularly imprinted magnetic solid-phase extraction materials based on magnetic carbon nanotubes and their application for the determination of gatifloxacin in serum samples coupled with high performance liquid chromatography. Journal of Chromatography A. 2013;**1274**:44-53

[69] Zhang W, Zhang Z, Zhang Y. The application of carbon nanotubes in target drug delivery systems for cancer therapies. Nanoscale Research Letters. 2011;**6**:555-577

[70] Pham-Huy LA, He H, Pham-Huy C. Free radicals, antioxidants in disease and health. International Journal of Biomedical Sciences. 2008;**4**(2):89-96

[71] Yang W, Thordarson P, Gooding JJ, Ringer SP, Braet F. Carbon nanotubes for biological and biomedical applications. Nanotechnology. 2007;**18**:12

[72] Madani SY, Naderi N, Dissanayake O, Tan A, Seifalian AM. A new era of cancer treatment: Carbon nanotubes as drug delivery tools. International Journal of Nanomedicine. 2011;**6**:2963-2979

[73] Jiang L, Liu T, He H, et al. Adsorption behavior of pazufloxacin mesilate on amino-functionalized carbon nanotubes. Journal of Nanoscience and Nanotechnology. 2012;**12**:1-9

[74] Liao H, Paratala B, Sitharaman B, Wang Y. Applications of carbon nanotubes in biomedical studies. Methods in Molecular Biology. 2011;**726**:223-241

[75] MacDonald RA, Laurenzi BF, Viswanathan G, Ajayan PM, Stegemann JP. Collagen-carbon nanotube composite materials as scaffolds in tissue engineering. Journal of Biomedical Materials Research A. 2005;**74**(3):489-496

[76] Zhang Y, Bai Y, Yan B. Functionalized carbon nanotubes for potential medicinal applications. Drug Discovery Today. 2010;**15**(11-12):428-435

[77] Wang J. Carbon-nanotube based electrochemical biosensors: A review. Electroanalysis. 2005;**17**(1):7-14

[78] Yang Z, Zhang Y, Yang Y, et al. Pharmacological and toxicological target organelles and safe use of single-walled carbon nanotubes as drug carriers in treating Alzheimer disease. Nanomedicine. 2010;**6**(3):427-441

[79] Zhou Z, Wang X, Faraji S, Bradford PD, Li Q, Zhu Y. Mechanical and electrical properties of aligned carbon nanotube/carbon matrix composites. Carbon. 2014;**75**:307-313

[80] Mirri F, Ma AWK, Hsu TT, Behabtu N, Eichmann SL, Young CC, et al. High-performance carbon nanotube transparent conductive films by scalable dip coating. ACS Nano. 2012;**6**(11):9737-9744

[81] Duesberg GS, Graham AP, Liebau M, Seidel R, Unger E, et al. Large-scale integration of carbon nanotubes into silicon-based microelectronics. Proceedings of SPIE, Nanotechnology. 2003;**5118**:125

[82] Aissa B, Nedil M, Habib MA, Abdul-Hafidh EH, Rosei F. High-performance thin-film-transistors based on semiconducting-enriched single-walled carbon nanotubes processed by electrical-breakdown strategy. Applied Surface Science. 2015;**328**:349-355

[83] Wang F, Kozawa D, Miyauchi Y, Hiraoka K, Mouri S, Ohno Y, et al. Considerably improved photovoltaic performance of carbon nanotube-based solar cells using metal oxide layers. Nature Communications. 2015. Article Number 6305

[84] Adeniran B, Mokaya R. Low temperature synthesized carbon nanotube superstructures with superior CO_2 and hydrogen storage capacity. Journal of Materials Chemistry A. 2015;**3**:5148-5161

[85] Li H, Liu W, Cassell AM, Kreupl F, Banerjee K. Low-resistivity long-length horizontal carbon nanotube bundles for interconnect applications—Part I: Process development. IEEE Transactions on Electron Devices. 2013;**60**(9):2862-2869

[86] Cai L, Wang C. Carbon nanotube flexible and stretchable electronics. Nanoscale Research Letters. 2015;**10**:320

[87] Lin X, Lin, Chen Z, Feng C, Liang L, Bai Z-Q, et al. Flexible, stretchable, transparent carbon nanotube thin film loudspeakers. Nano Letters. 2008;**8**(12):4539-4545

[88] Ong YT, Ahmad AL, Zein SHS, Tan SH. A review on carbon nanotubes in an environmental protection and green engineering perspective. Brazilian Journal of Chemical Engineering. 2010;**27**(2)

[89] Xia N, Gao Y. Carbon nanostructures for development of acetylcholinesterase electrochemical biosensors for determination of pesticides. International Journal of Electrochemical Science. 2015;**10**:713-724

[90] Makowski T, Kowalczyk D, Fortuniak W, Jeziorska D, Brzezinski S, Tracz A. Superhydrophobic properties of cotton woven fabrics with conducting 3D networks of multiwall carbon nanotubes, MWCNTs. Cellulose. 2014;**21**:4659-4670

[91] Kang JW, Song KO, Hwang HJ, Jiang Q. Nanotube oscillator based on a short single-walled carbon nanotube bundle. Nanotechnology. 2006;**17**(9):2250

[92] Wei J, Zhu H, Wu D, Wei B. Carbon nanotube filaments in household light bulbs. Applied Physics Letters. 2004;**9**(24):4869-4871

[93] Kyatskaya S, Mascarós JRG, Bogani L, Hennrich F, Kappes M, Wernsdorfer W, et al. Anchoring of rare-earth-based single-molecule magnets on single-walled carbon nanotubes. Journal of American Chemical Society. 2009;**131**(42):15143-15151

[94] Maksimenko SA, Slepyan GY, Nemilentsau AM, Shuba MV. Carbon nanotube antenna: Far-field, near-field and thermal-noise properties. Physica E: Low-dimensional Systems and Nanostructures. 2008;**40**(7):2360-2364

[95] Yildiz O, Bradford PD. Aligned carbon nanotube sheet high efficiency particulate air filters. Carbon. 2013;**64**:295-304

[96] Das R, Ali ME, Hamid SBA, Ramakrishna S, Chowdhury ZZ. Carbon nanotube membranes for water purification: A bright future in water desalination. Desalination. 2014;**336**:97-109

[97] Yu D, Liu H, Peng LM, Wang S. Flexible light-emitting devices based on chirality-sorted semiconducting carbon nanotube films. ACS Applied Materials & Interfaces. 2015;**7**(6):3462-3467

Chapter 12

Applications of Cadmium Telluride (CdTe) in Nanotechnology

Akeel M. Kadim

Abstract

Cadmium telluride quantum dots (CdTe QDs) were prepared by chemical reaction and used to fabricate electroluminescence quantum dot hybrid junction device. QD-LED was fabricated using TPD: PMMA/CdTe/Alq_3 device which synthesized by phase segregation method. The hybrid white light-emitting devices consist of three layers deposited successively on the ITO glass substrate; the first layer was of tetra-phenyl diaminobiphenyl (TPD) polymer mixed with polymethyl methacrylate (PMMA) polymers, while the second layer was 0.5 wt% of the (CdTe) QDs for hybrid device, whereas the third layer was tris (8-hydroxyquinoline) aluminum (Alq_3). The organic light-emitting device (OLED) was considered by room temperature photoluminescence (PL) and electroluminescence (EL). Current-voltage (I-V) characteristics indicate that the output current is good compared to the few voltage (6 V) used which gives good results to generate white light. The electroluminescence (EL) spectrum of hybrid device shows a wide emission band covering the range 350–700 nm. The emissions causing this white luminescence were identified depending on the chromaticity coordinates (CIE 1931): x = 0.32, y = 0.33. The correlated color temperature (CCT) was found to be about 5886 K. Fabrication of EL devices from semiconductor material (CdTe QDs) between two layers, organic polymer (TPD) and organic molecules (Alq_3), was effective in white light generation. The recombination processes and I-V characteristics give rise to the output current which is good compared to the few voltages used which give good results to generate light.

Keywords: CdTe, quantum dots, organic device, quantum hybrid device

1. Introduction

For the past several years, CdTe quantum dots (QDs) have been reconsidered extensively because of their potential for optoelectronic and biological nanoapplications. The unique advantage of colloidal QDs is their size-dependent physical and optical properties such as the energy band gap, narrow emission with small full width at half maximum, broad spectral photo response from ultraviolet to infrared regions, and their compatibility with solution processing [1].

QDs success more significance now a day due to their indicating of nanotechnology applications in the field of laser, bio-imaging, LED, and sensors [1, 2]. QDs materials can illustrate tunable photoluminescent property by changing the particle size. In particular, QDs-based light-emitting diodes (QD-LEDs) have been below the global attention as a developing technology for next-generation displays or solid-state lighting. A huge improvement has been created in the enhancement of high-performance QD-LEDs of which brightness and efficiency are comparable to those of OLEDs [2].

Among several QDs, cadmium telluride (CdTe) QDs have been significantly used in work and biomedical applications owing to their tunable photoluminescence inside the visible range once excited by a single excitation wavelength. For evidence, CdTe QDs are expected to be possible probes in the bio-imaging of living cells as of their many benefits for example higher photo stability, more controllable and narrower emission bands, and higher quantum fabricate in relationship with conformist fluorescent dyes [3].

The fast enhancement in synthetic techniques has certified producing semiconductor nanoparticles of narrow sizes and of more than a few shapes. The phenomenon of attractive has a great influence of nano materials of the construction of these materials and their optical properties [4]. For existence, magic-sized CdTe has a very broad so-called white light emission with extensive emission quantum produce, which is very different from the very narrow band gap emission detected from typical semiconductor QDs [3, 4].

A leading difference to OLEDs is that the active layer covers ionic components in addition to the light-emitting species (polymers or transition metal complexes). These ionic kinds start moving under applied voltage and so enable charge injection into the light-emitting component [5, 6].

So, the charge transporter injection and transport in this type of devices is ongoing by the movement of the ionic types, the response and turn-on time of OLEDs which is acceptably long, ranging usually from subseconds to hours, depending on the ionic conductivity of the light-emitting layer [7, 8].

The hybrid device was fabricated from three layers: the first layer is the TPD polymer, the second is CdTe QDs, and the third is Alq_3; all these layers are on ITO substrate. It has a significant effect in application of nanotechnology to getting a clear and efficient high-intensity of white light generation [8, 9]. The emission of EL of the hybrid devices showed luminescence of white light with high intensity and good efficiency, using a few voltages [9].

2. Synthesis the CdTe quantum dots

All materials used in this work were supplied from Fluka Company without further purification; TPD is a hole-transport molecule, having maximum absorption wavelength at 351 nm and emission wavelength at 391 nm, while Alq_3 is an electron transport molecule, having maximum absorption wavelength at 392 nm and emission wavelength at 519 nm. The cadmium telluride CdTe QDs were made by combination two chemical solutions of molarities 0.02 M. The first solution was arranged by dissolving 0.092 g of $CdCl_2$ in 50 ml distilled water, while the second solution was succeeded by dissolving 0.033 g from sodium telluride Na_2Te in 50 ml distilled water. The two solutions were mixed at 1:1 mole ratio in a three-neck flask and left on magnetic stirrer at temperature of around 80°C; then, ammonium hydroxide (NH_4OH) was added drop by drop to the solution, having three values of pH 10 with continuous flowing of argon gas for about 1 h, till the color changed to light green.

Fabrication process of the hybrid junction devices can be summarized. It consists of three coatings added sequentially on the ITO glass substrate by phase

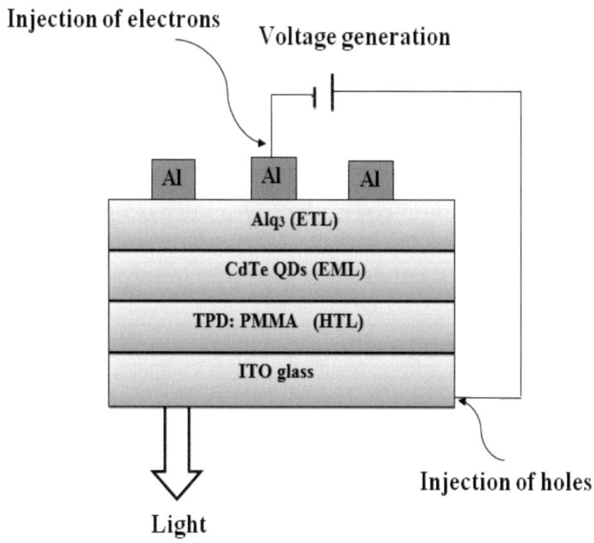

Figure 1.
Structure of TPD: PMMA/CdTe/Alq$_3$ hybrid junction devices.

segregation method using spin coating at 2000 r.p.m. for about 10–15 s for each coat. The first layer was of TPD mixed with PMMA in ratio 1:1. PMMA was used to prevent cracks in the film, leading to the increased conductivity of TPD, while the second layer was 0.5%wt CdTe QDs. After deposition, each layer is dried at a temperature of 60°C to get a layer well rolled on the film. Completely deposition Alq$_3$ coat above the CdTe QDs film. The thicknesses of the films have been measured by interference method whereas the thicknesses of TPD: PMMA and Alq$_3$ layers were 600 and 540 nm, respectively; while the thickness of CdTe QDs layers was 550 nm, whereas the thickness and resistance of ITO coat were 150 nm and 10 Ω individually. Then, aluminum cathode is added on the hybrid film. The hybrid junction device is thus sandwiched between the ITO and aluminum electrodes. The ratio of TPD: PMMA/CdTe QDs/Alq$_3$ was taken to be 1 ml:1 ml/0.5%wt/1 ml separately (see **Figure 1**).

Conducting layer deposit is an organic contain of TPD indicate a hole- carrying layer (HTL), while the emissive cover of QDs be an electron-injected layer (EML) then Alq$_3$ illustrate an electron transporting layer (ETL). The Alq$_3$ molecular orbitals experienced position on states surfaces of semiconductors materials. These molecules orbitals are trapped to the Fermi level in QDs materials because of that injunction of electrons charge assignment and substrate work function main to trapped level (TL) placement within the high occupied molecular orbital (HOMO) and low un occupied molecular orbital (LUMO) gap. Successively a potential is applied, the injected positive and negative carrier's charges recombine process in the emissive layer to produce an electroluminescence light. The transport electrons from the cathode (Al) and produces light in reaction to an electric current as appear in **Figure 1**. Recognizable anode films are thin films of optically transparent and electrically conductive material.

3. Characterization

Cadmium telluride QDs were characterized by OPTIMA SP- 3000 UV-Vis spectrometer in the spectral range 200–1100 nm and photoluminescence spectrum was measured by SL 174 spectrofluorometer covering a range 300–900 nm. The scanning electron microscope (SEM) was recorded by VEGA3 TESCAN, mode SE from TESCAN ORSAY HOLDING. The I-V measurements for the TPD: PMMA/CdTe/Alq_3 hybrid device hybrid junction was approved out by Keithley digital electrometer 616 and D.C. power supply to estimate the EL device. The EL spectrum of the EL devices was measured at room temperature using fiber optic spectrometer CCS Series by THORLABS Company (Germany) with a covering range of 200–1000 nm and resolution of 1 nm.

4. Optical properties of CdTe QDs

The absorption spectrum of the CdTe QDs is shown in **Figure 2**.

It is can be noticed from **Figure 2** that the absorbance spectrum of CdTe QDs illustrates high absorbance in the visible range 350–450 nm and extremely decreases till about 470 nm where there is no absorbance [10]. The spectrum shows high absorbance of CdTe QDs in the ultraviolet region, where the absorption spectrum of the CdTe is characterized by convexity or peaks resulting from a nanostructure formation of the material.

The photoluminescence (PL) spectrum of the CdTe QDs is revealed in **Figure 3**.

Figure 3 shows the PL of the colloidal CdTe QDs, which shows that the band edge transmission is centered at 550 nm and other peaks represent the surface states

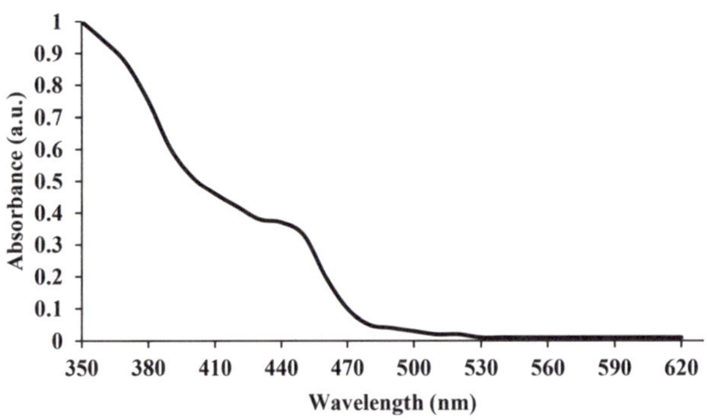

Figure 2.
Absorption spectrum of CdTe QDs.

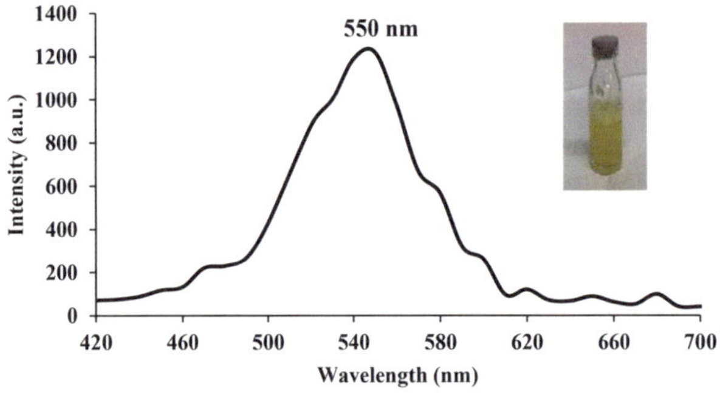

Figure 3.
Photoluminescence (PL) spectrum of CdTe QDs.

[10, 11]. This value refers to the band-to-band transition in CdTe QDs, where the emission spectrum of the CdTe is characterized by high intensity due to the presence of surface states that cause an increase in the intensity and efficiency of the material that has wide applications in nanotechnology applications [11]. The energy gap calculated from PL according to the relation ($E = 1240/\lambda$ (nm)) was found to be about 2.25 eV for CdTe QDs.

5. Morphological properties of CdTe QDs

5.1 Scanning electron microscope (SEM)

The surface morphology of the arranged CdTe QDs was considered by the SEM of 60K× magnifications, as revealed in **Figure 4**. The SEM images of the QDs films give a good sign for construction of the CdTe QDs. The average grain size governed from SEM is about 10 nm by knowing the diameter of the QD and dividing it by the amount of magnification used. **Figure 4** reveals that the shape of formed QDs is approximately spherical, while the image shows aggregation of QDs in the range of 100 nm.

Figure 4.
Scanning electron microscope (SEM) of CdTe QDs.

6. Electrical properties of CdTe QDs

Figure 5 shows the I-V characteristics of the hybrid junction devices obtained using the TPD: PMMA/CdTe/Alq_3. **Figure 5** reveals that the rectification behavior with a turn-on voltage in general at bias voltage at 6 V, while light emissions was found at current levels of near 0.02–0.7 mA.

The I-V appearances of the hybrid device indication exponential increase in current because of reduction in the depletion layer width at the border. In the forward bias, the conduction band barrier will shrunk due to the exponential distribution of electrons and holes in the conduction and valence bands and thus the diffusion current running through the hybrid junction increases exponentially with increasing forward bias. The drift current curving in the opposite direction does not depend on the potential barrier height and will improve the electron flow from the Alq_3 to n-(QDs) and holes from the p-(TPD) to the n-(QDs) [12]. The succeeding recombination would contribute to increase in the forward bias current flow with few voltages at 6 V for TPD: PMMA/CdTe/Alq_3.

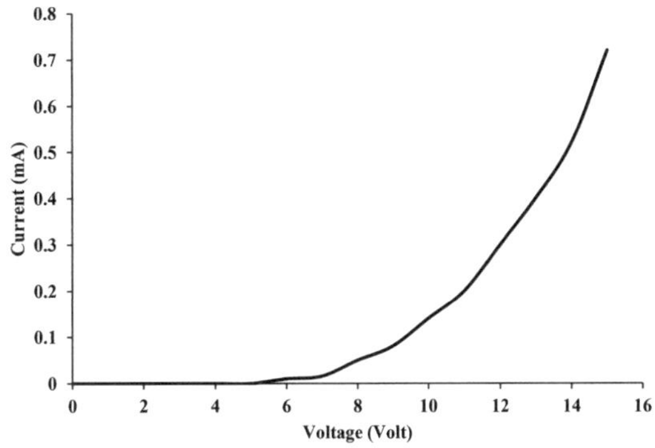

Figure 5.
Current-Voltage (I-V)characteristics of TPD:PMMA/CdTe/Alq$_3$ hybrid EL devices.

7. Electroluminesces properties of CdTe QDs

The EL measurements under forward bias voltages of 6 V represent the upper limit for the light which has been obtained experimentally from the TPD: PMMA/0.5%wt CdTe QDs/Alq$_3$ hybrid junction devices; light emission was carried out using a photomultiplier detector at room temperature.

Figure 6 represents the emission spectrum studied by CIE 1931 chromaticity diagram, which indicates the white light generated at forward bias voltage of 6 V. It is clear from the figure that the peaks at 460, 540, and 610 nm of TPD: PMMA/ CdTe/Alq$_3$ and the other peaks are due to defect states. The mechanism transport of carrier in the hybrid junction device (QDs-OLEDs) is that the TPD performs as the hole transporting material and it contributes to increase the intensity emission of OLEDs [13]. Holes are injected from the ITO anode into the high occupied molecular orbital (HOMO) of the TPD material and carried to the valance band; electrons are injected from the aluminum (Al) cathode into the conduction band. These electrons are injected once more in Alq3 layer, which is used as an electron transfer material and emitting layer. The number and the mobility of electrons will be improved and consequently are receiving high mobility of electrons affecting the CdTe QDs layer [14].

Therefore, holes and electrons from the excitons in the QDs that recombine radiatively are called band-to-band recombination. The holes and electrons recombination achieved defects sates to emission light in changed wavelengths, this recombination processes complete defects names the Shockley-Read-Hall recombination. The maximum emission intensity increases were creating in the case of layers covering 0.5wt%

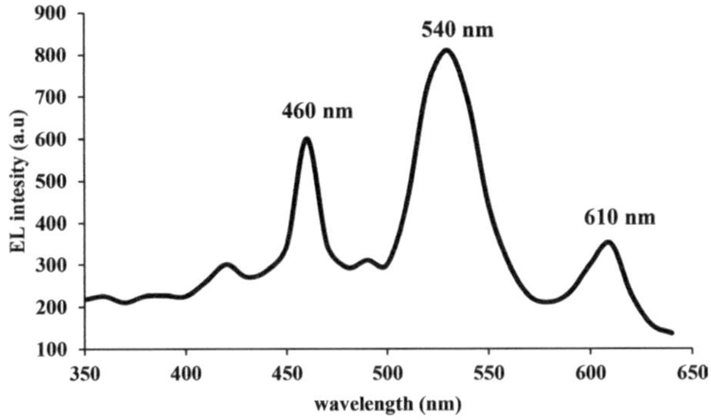

Figure 6.
Electroluminescence of TPD: PMMA/CdTe/Alq$_3$ hybrid EL device.

of QDs and Alq_3 layer when compared with that without Alq_3 layer. Addition of Alq_3 organic molecules layer causes in generation of Förster energy between the LUMO epitomized by (Alq_3) and HOMO signified by (ITO), which indicates increase in the efficiency of hybrid devices [15]. From **Figure 6**, it can be perceived the occurrence of peaks spectrum at blue, green, and red, which in turn contributes to white light when mixed. The energy unconfined from the recombination of the charge transfer excitons being resonantly transported to the proximal electrons in conduction band of the QDs concluded an Auger process to produce electrons with sufficiently high energy to inject into the lower unoccupied molecular orbital (LUMO) of Alq_3. These electrons then radioactively recombine with holes in the HOMO of the polymer, causing emission of photons with energy equal to the HOMO-LUMO gap of the TPD [15, 16].

The white light generated by hybrid devices, or several other light sources for general lighting, should have a good white color in stability to show all the colors of illuminated objects suitably. As the color of light is expressed by the CIE colorimetry system [17], the spectrum of a given light is weighted by the XYZ color matching functions. The x and y on CIE system are located from X, Y, and Z:

$$x = \frac{X + Y + Z}{X} \tag{1}$$

$$y = \frac{X + Y + Z}{Y}$$

The correlated color temperature (CCT) for any white light can be analyzed by using McCamy's approximation algorithm to estimate the CCT from the x, y chromaticity coordinates as in Eq. (2):

$$\text{CCT} = -449\,n^3 + 3525\,n^2 - 6823n + 5520.3\ldots\ldots \tag{2}$$

where $n = \frac{x - 0.3320}{y - 0.1858}$.

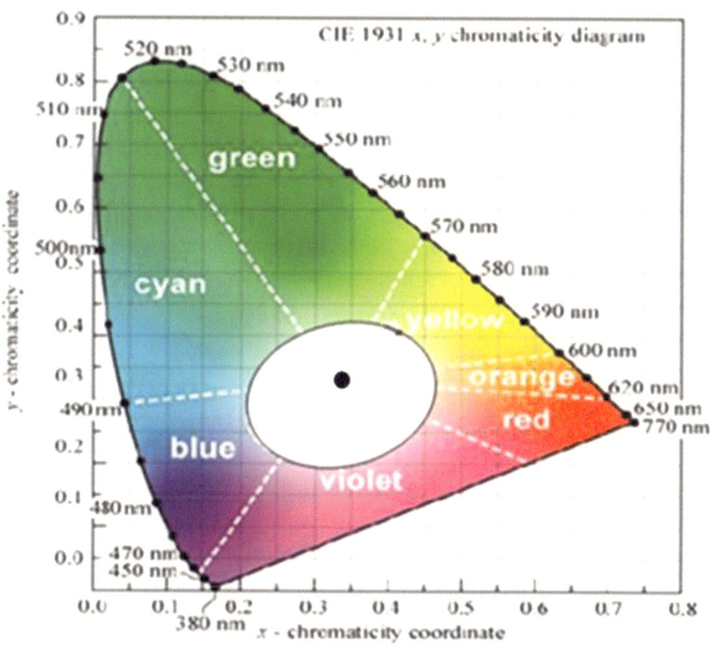

Figure 7.
Tristimulus coordinates of hybrid device on the chromaticity diagram.

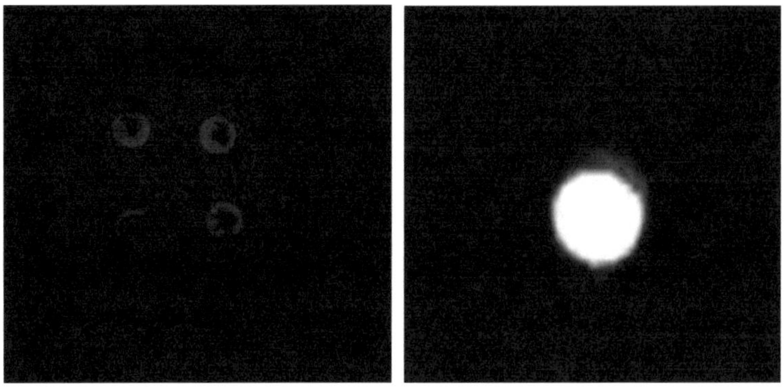

Figure 8.
A photographic plate of the white light generation of TPD: PMMA/CdTe/Alq$_3$.

Figure 7 shows the CIE 1931 chromaticity diagram (x = 0.32, y = 0.33). The output light of the emission lines registered in the EL spectrum is in the white light region confirmed by the photograph of the white light emitted from the hybrid devices in **Figure 8**. It is clear that the intensity of the output light is very high and clear.

The cause of emission of this white light was identified depending on chromaticity coordinates. The values of chromaticity coordinates show that the hybrid device of TPD: PMMA/0.5%wtCdTe/Alq$_3$ has a high correlated color temperature of about 5886 K. This means that during high temperature, the light site will be heading toward the center of the white light, which in turn gives a high efficiency of the devices.

8. Summary

Limited confinement size of CdTe QDs is considered by chemical effect which was very beneficial since they have many defects. These defects sates can be used in many applications for occurrence of white light generation in QDs-OLEDs. The producer of white light properties with high efficiency using confinement effect makes a large energy gap and hence, the direction of the light sites be toward the center of white light color. The succeeding recombination processes would give rise to the forward bias current flow. So growth in the forward current under high bias could be the amplification of the good contact between the Al electrode and QDs layer. The output current from I-V forms is good compared to the few voltages used which give good results to develop a production of white light. Fabrication of EL device from semiconductors material (CdTe QDs) with hole injection organic polymer (TPD) and electron injection (Alq3) was effective in the intensity and efficiency of white light generated and can be the color of emerged clear light.

Author details

Akeel M. Kadim
Biomedical Engineering Department, College of Engineering, University of Warith Al Anbiyaa, Iraq

*Address all correspondence to: ph.d.akeel@gmail.com; akeel.kadim@uowa.edu.iq

References

[1] Frohleiks J, Wepfer S, Kelestemur Y, Demir HV, Bacher G, Nannen E. Quantum dot light emitting electrochemical cell hybrid device and mechanism of its operation. ACS Applied Materials & Interfaces. 2016;**8**:24692-24698. DOI: 10.1021/acsami.6b06833

[2] Dai J, Ji Y, Xu CX, Sun XW, Leck KS, Ju ZG. White light emission from CdTe quantum dots decorated n-ZnO nanorods/p-GaN light-emitting diodes. Applied Physics Letters. 2011;**99**:063112

[3] Tan F, Qu S, Zhang W, Wang Z. Hybrid morphology dependence of CdTe:CdSe bulk heterojunction solar cells. Nanoscale Research Letters. 2014;**9**:1-8

[4] Liu R. Hybrid organic/inorganic nanocomposites for photovoltaic cells. Materials (Basel). 2014;**7**:2747-2771. Available from: https://www.ncbi.nlm.nih.gov/pmc/articles/PMC5453346

[5] Chen Z, Zhou Y, Zhang T, Bu X, Sheng X. Preparation and characterization of optically active polyacetylene and CdTe quantum dots composites with low infrared emissivity. Journal of Inorganic and Organometallic Polymers. 2014;**24**:591-599

[6] Dagtepe P, Chikan V, Jasinski J, Leppert VJ. Quantized growth of CdTe quantum dots; observation of magic-sized CdTe quantum dots. Journal of Physical Chemistry C. 2007;**111**:14977-14983

[7] Lee JH, Shin H, Kim J, Kim KH, Kim JJ. Exciplex-forming co-host-based red phosphorescent organic light-emitting diodes with long operational stability and high efficiency. ACS Applied Materials and Interfaces. 2017;**9**:3277-3281. DOI: 10.1021/acsami.6b14438

[8] Ge Z et al. Improved performance of silicon nanowire/cadmium telluride quantum dots/organic hybrid solar cells. Applied Surface Science. 2014;**334**:15-18. DOI: 10.1016/j.apsusc.2014.07.063

[9] Jana B, Ghosh A, Maiti S, Bain D, Banerjee S, Ghosh HN, et al. Size of CdTe quantum dots controls the hole transfer rate in CdTe quantum dots–MEHPPV polymer nanoparticle hybrid. Journal of Physical Chemistry C. 2016;**120**:25142-25150. DOI: 10.1021/acs.jpcc.6b09986

[10] Saikia D. Optical properties of CdTe quantum dots synthesized at different pH. International Journal of Interdisciplinary Research in Science Society and Culture (IJIRSSC). 2016;**2**:2395-4335

[11] Dutta P, Beaulac R. Photoluminescence quenching of colloidal CdSe and CdTe quantum dots by nitroxide free radicals. Chemistry of Materials. 2016;**28**:1076-1084. DOI: 10.1021/acs.chemmater.5b04423

[12] Kadim AM. Zinc selenide quantum dots light emitting devices (ZnSe QDs-LEDs) with different organic polymers. Nano Hybrids and Composites. 2017;**18**:11-19

[13] Kim HH, Park J, Han IK, Won SO, Park C, Hwang DK, et al. Emissive CdTe/ZnO/GO quasi core-shell-shell hybrid quantum dots for white light emitting diodes. Nanoscale. 2016;**8**:19737-19743

[14] Caixia Y, Yifeng T. The study on a CdTe quantum dot electrochemiluminescent sensor. In: International Conference on Enabling Science and Nanotechnology (ESciNano); 2010

[15] Thuy UTD, Chae W, Yang W, Liem NQ. Enhanced fluorescence properties of Type-I and Type-II CdTe/CdS quantum dots using porous silver membrane. Optical Materials. 2017;**66**:611-615

[16] Schanda J. Colorimetry: Understanding the CIE System. 1st ed. USA: Wiley-Interscience (US); 2008

[17] McCamy CS. Correlated color temperature as an explicit function of chromaticity coordinates. Color Research and Application. 2007;**17**:142-144